M000275854

ADVANCED EUCLIDEAN
GEOMETRY

EXCURSIONS FOR SECONDARY
TEACHERS AND STUDENTS

ALFRED S. POSAMENTIER
The City College, The City University of New York

Key College Publishing
Innovators in Higher Education
www.keycollege.com

Key College Publishing was founded in 1999 as a division of Key Curriculum Press in cooperation with Springer-Verlag New York, Inc. We publish innovative texts and courseware for the undergraduate curriculum in mathematics and statistics, as well as mathematics and statistics education. For more information, visit us at www.keycollege.com.

Key College Publishing
1150 65th Street
Emeryville, CA 94608
info@keycollege.com
www.keycollege.com
(510) 595-7000

Alfred S. Posamentier
The City College, The City University of
    New York
New York, NY 10031

Library of Congress Cataloging-in-Publication Data
Posamentier, Alfred S.
    Advanced Euclidean geometry : excursions for secondary
teachers and students / by Alfred S. Posamentier.
        p. cm.
        ISBN 1-930190-85-9 (pbk.)
        1. Geometry, Plane. I. Title.
QA45 .P68 2002
516.22—dc21   2001057968

© 2002 by Key College Publishing

This work is protected by federal copyright law and international treaty. The book may not be translated or copied in whole or in part without the permission of the publisher, except for brief excerpts in connection with reviews or scholarly analysis. Use in connection with any form of information storage and retrieval, electronic adaptation, or computer software, or by similar or dissimilar methodology now known or hereafter developed, is forbidden.

The use of general descriptive names, trade names, trademarks, etc., in this publication, even if the former are not specially identified, is not to be taken as a sign that such names, as understood by the Trade Marks and Merchandise Marks Act, may accordingly be used freely by anyone. Where those designations appear in this book and the publisher was aware of a trademark claim, the designations follow the capitalization style used by the manufacturer.

®The Geometer's Sketchpad is a registered trademark of Key Curriculum Press. ™Sketchpad is a trademark of Key Curriculum Press. All other registered trademarks and trademarks in this book are the property of their respective holders.

Advanced Euclidean Geometry CD-ROM
Key College Publishing guarantees that the CD-ROM that accompanies this book is free of defects in materials and workmanship. A defective disk will be replaced free of charge if returned within 90 days of the purchase date. After 90 days, there is a $10.00 replacement fee.

Project Editor:	Cortney Bruggink
Production Editor:	Christine Osborne
Copy Editor:	Mary Roybal
Production Director:	Diana Jean Parks
Text Designer:	Garry Harman/The ArtWorks
Compositor:	GTS Graphics
Illustrator:	Lineworks, Inc.
Technical Artist:	Jan Siwanowicz
Photo Researcher:	Margee Robinson
Cover Designer:	Caroline Ayres
Photo Credits:	Cover: Charles O'Rear/Corbis; 217 (*l*): Eye Ubiquitous/Corbis; 217 (*r*): Archivo Iconografico, S. A./Corbis
Prepress:	GTS Graphics
Printer:	Data Reproductions Corporation
Executive Editor:	Richard Bonacci
General Manager:	Mike Simpson
Publisher:	Steven Rasmussen

Printed in the United States of America
10 9 8 7 6 5 4 3 2   06 05 04

Contents

PREFACE

Until the nineteenth century, it was thought that all that was significant about the geometry of the triangle and the circle had been discovered by Euclid and his predecessors. However, during the nineteenth century, a plethora of articles appeared that expanded this field enormously. Many additional relationships were discovered that brought new life to the field of Euclidean geometry. This book presents the highlights of these newer discoveries in a reader-friendly format. In short, this book is designed to provide an extended view of Euclidean geometry in order to expand the background of the secondary school mathematics teacher.

Over the past three hundred years, many textbooks have been written to present Euclid's *Elements* to a school audience. The most notable of these are Robert Simson's *Elements of Euclid,* which first appeared in 1756, and Adrien-Marie Legendre's *Elements de geometrie,* which was published (in French) in 1794. An English version of Legendre's text was revised in 1828 by Charles Davies, a West Point professor. "Davies' Legendre," as it is popularly referred to, was one of the most widely used American mathematics textbooks of the nineteenth century and perhaps has had the greatest influence on our present-day high school geometry course of any text.

Legendre's geometry did not state the theorems in general terms. Rather, it employed diagrams to demonstrate the various propositions. This departure from Euclid was corrected by Davies, who provided a general statement of a proposition followed by an explanation and an accompanying diagram. This book uses both approaches interchangeably, as appropriate.

Our study of geometry—advanced Euclidean geometry—begins where the high school geometry course (still for the most part fashioned after Davies' Legendre) leaves off. This book does not attempt to provide an exhaustive study of the entire field of these advanced topics, which would be impossible in one small book. Instead we focus our attention on subjects that are of interest to those who have mastered the high school geometry course, have a genuine desire or need to extend their study of mathematics, and will appreciate the beauty that lies in the study of advanced Euclidean geometry.

A unique feature of this book is the inclusion of interactive geometric figures provided on a CD-ROM using The Geometer's Sketchpad®, software. All too often, geometry is presented in a static form in which the true and deeper meaning of a theorem does not get the true exposure it should. The reader is encouraged, whenever an interactive geometric figure is indicated by the CD-ROM icon (shown at left), to go to the computer and explore the figure by distorting it and observing the constancy being established.

To truly understand a subject and to teach it well, one must know more about the subject than the material being taught. The material in this book has been tested and evaluated during more than twenty-five years of use with numerous classes of secondary school teachers at The City College, The City University of New York. Many valuable suggestions have been received and incorporated into this book.

A number of people provided technical support, for which I am profoundly grateful. For creating The Geometer's Sketchpad, drawings (both static and interactive), often in most ingenious ways, much credit must go to Jan Siwanowicz. David Linker proofread the entire manuscript. In helping develop the *Instructor Resources,* a group of highly talented students prepared some wonderful solutions to the exercises in the book. These students included Kamaldeep Gandhi, Seth Kleinerman, Leo Nguyen, Oana Pascu, Peter Ruse, and Jan Siwanowicz. The technical typing for the entire manuscript was done in stellar form by Sandra Finken. Above all, I wish to thank the hundreds of students (high school math teachers in their own right) who have used part of this book over the past several years for their valuable comments about its contents. These comments kept me properly focused!

Alfred S. Posamentier

INTRODUCTION

This book undertakes topics that are beyond the scope of the typical high school geometry course, but it treats the topics using elementary methods and nomenclature. Thus, the book may be easily understood by interested high school students even though it is aimed particularly at the in-service or pre-service secondary school mathematics teacher. The use of familiar language means that readers do not have to learn entirely new concepts and skills, only new uses for their previous knowledge base. Readers are provided an opportunity to extend their knowledge of Euclidean geometry in a style to which they are accustomed. The book also provides secondary teachers with a wealth of ideas to enrich their instructional program.

Chapter 1 reviews the essentials of the high school geometry course. To focus a critical eye on this material, we inspect some fallacies in elementary Euclidean geometry. The discovery of these fallacies sharpens geometric awareness. Chapters 2 and 3, linked by the concept of duality, deal with the often-neglected topics of concurrency and collinearity. Theorems, rather difficult to prove in the high school geometry course, will now be much easier to prove. Moreover, experimenting with the figures on the CD-ROM will show that what is stated as a theorem proves to be true as the diagram is manipulated to demonstrate a multitude of possible cases. This is what a proof typically establishes. Facility with concurrency and collinearity enables a simple development of some other interesting theorems explored in these chapters.

In the next two chapters, our attention turns to the triangle. Chapter 4 begins our discussion by looking at some rather unusual points in a triangle. Chapter 5 introduces properties of various interior segments of triangles (often referred to as Cevians), including angle bisectors and medians. Other triangle properties not previously encountered are also considered here.

The treatment of quadrilaterals in high school is limited to the special quadrilaterals: parallelogram, rhombus, rectangle, square, and trapezoid. Our study of quadrilaterals in Chapter 6 assumes a knowledge of the properties of these special quadrilaterals. We begin with the general convex quadrilateral and eventually turn to the cyclic (or inscribed) quadrilateral. With the aid of Ptolemy's theorem, we establish many interesting geometric relationships.

The only two circles associated with a triangle in the high school geometry course are the circumscribed and inscribed circles. While the inscribed circle is tangent to the three sides of the triangle and lies *inside* the triangle, its analog, the escribed circle (or excircle) of a triangle, is also tangent to the three sides of the triangle (or their extensions) but lies *outside* the triangle. The inscribed circle and the three escribed circles of a triangle are known as the equicircles of the triangle. Chapter 7 explores some of the many relationships that involve the equicircles of a triangle.

A popular topic in advanced Euclidean geometry, with a host of surprising properties, is the *nine-point circle.* Midway through the investigation of the nine-point circle in Chapter 8, we digress to study some properties of the altitudes of a triangle and their associated orthic triangle that will permit us to develop further properties of the nine-point circle.

One of the most creative problem-solving challenges in geometry can be found in constructing triangles given the measures of three parts of the triangle, such as the lengths of its three medians, the lengths of its three altitudes, or the measures of two of its angles and the length of the included side. Such construction problems are presented in Chapter 9, with many illustrative examples and plenty of exercises. Requiring nothing more advanced than a knowledge of high school geometry, these construction problems offer ample opportunity to challenge even the best geometricians!

The "problem of Apollonius" has intrigued generations of mathematicians. It is presented in Chapter 10 as an application of circle constructions that follow certain requirements, such as passing through a given point and/or tangent to a given line and/or tangent to a given circle. While some of these constructions may be rather trivial, others are extremely challenging and were the focus of mathematicians in the seventeenth and eighteenth centuries.

The mission of the final chapter is to demonstrate a connection between Euclidean geometry and other branches of mathematics. This is done through the study of the golden section and Fibonacci numbers. Chapter 11 merely scratches the surface of an extremely rich topic. The extended exercise section should serve as a springboard for further investigation.

ABOUT THE AUTHOR

Alfred S. Posamentier is Professor of Mathematics Education and Dean of the School of Education of The City College, The City University of New York. He is the author and coauthor of many mathematics books for in-service and future teachers and for secondary school students. His publications concentrate on the teaching of mathematics, problem solving, and enrichment topics from advanced Euclidean geometry. This book is the outgrowth of many years of experience working with in-service high school teachers to enrich their teaching of geometry.

Dr. Posamentier received his Ph.D. in mathematics education from Fordham University. After six years as a high school teacher, Dr. Posamentier joined the faculty of The City College, and soon thereafter he began to develop in-service courses for secondary school mathematics teachers. He is currently involved in working with mathematics teachers, both in the United States and internationally, to help them better understand the ideas presented in this book so that they can comfortably incorporate them into their regular instructional program. He is an Honorary Fellow at the South Bank University in London, England. He has been a Fulbright Professor at the University of Vienna and a visiting professor at several European universities, including the Technical University of Vienna and the Humboldt University at Berlin.

Dr. Posamentier has been cited for his outstanding teaching, both in the United States and in Europe. He was named Educator of the Year (1993) by The City College Alumni Association. He was awarded the Grand Medal of Honor (1994) from the Federal Republic of Austria and the Medal of Distinction (1997) from the City of Vienna (Austria). He recently was awarded the title of University Professor for Austrian Universities (1999).

After more than thirty-two years on the faculty of CCNY, Dr. Posamentier still exudes an ever-increasing energy and enthusiasm for mathematics and mathematics education. In particular, his love for geometry can be felt in the pages of this book. By looking at some of the classical topics, all too often neglected or simply pushed out of the limelight for more "glitzy" topics, Dr. Posamentier hopes to bring this time-honored subject to the fore with the aid of the latest technology.

ELEMENTARY
EUCLIDEAN
GEOMETRY
REVISITED

REVIEW OF BASIC CONCEPTS OF GEOMETRY

Because the high school geometry course contains many theorems that are not easily remembered, we should take a brief look at some of the more important theorems. Our approach here, however, will differ from that used in your initial exposure to the theorems. We will consider the theorems according to their respective topics, not necessarily in the sequence originally presented but in a clear and concise fashion.

I. **Quadrilaterals**
    A. Methods of proving that a quadrilateral is a *parallelogram*
       To prove that a quadrilateral is a parallelogram, prove that:
       1. Both pairs of opposite sides are parallel.
       2. Both pairs of opposite sides are congruent.
       3. One pair of sides are both congruent and parallel.
       4. Both pairs of opposite angles are congruent.
       5. One pair of opposite angles are congruent and one pair of opposite sides are parallel.
       6. The diagonals bisect each other.

    B. Methods of proving that a quadrilateral is a *rectangle*
       To prove that a quadrilateral is a rectangle, prove that:
       1. It has four right angles.
       2. It is a parallelogram with one right angle.
       3. It is a parallelogram with congruent diagonals.

    C. Methods of proving that a quadrilateral is a *rhombus*
       To prove that a quadrilateral is a rhombus, prove that:
       1. It has four congruent sides.
       2. It is a parallelogram with two consecutive sides congruent.
       3. It is a parallelogram in which a diagonal bisects an angle of the parallelogram.
       4. It is a parallelogram with perpendicular diagonals.

    D. Methods of proving that a quadrilateral is a *square*
       To prove that a quadrilateral is a square, prove that:
       1. It is a rectangle with two consecutive sides congruent.
       2. It is a rectangle with a diagonal bisecting one of its angles.
       3. It is a rectangle with perpendicular diagonals.
       4. It is a rhombus with one right angle.
       5. It is a rhombus with congruent diagonals.

E. Methods of proving that a *trapezoid is isosceles*
  To prove that a trapezoid is isosceles, prove that:
    1. Its nonparallel sides are congruent.
    2. The base angles are congruent.
    3. The opposite angles are supplementary.
    4. Its diagonals are congruent.

  Note: We define a trapezoid as a quadrilateral with *exactly one* pair of opposite sides parallel. We do, however, note that some texts consider a trapezoid to be a quadrilateral with *at least one* pair of opposite sides parallel.

II. Midline of a Triangle

A. The *midline* of a triangle is the line segment joining the midpoints of two sides of the triangle.

B. The midline of a triangle is parallel to the third side of the triangle.

C. The midline of a triangle is half as long as the third side of the triangle.

D. If a line containing the midpoint of one side of a triangle is parallel to a second side of the triangle, then it also contains the midpoint of the third side of the triangle.

III. Similarity

A. When a line is parallel to one side of a triangle
    1. If a line parallel to one side of a triangle intersects the other two sides, then it divides these two sides proportionally.
    2. If a line divides two sides of a triangle proportionally, then the line is parallel to the remaining side of the triangle.

B. Proportionality involving *angle bisectors*
    1. An interior angle bisector of any triangle divides the side of the triangle opposite the angle into segments proportional to the adjacent sides. In Figure 1-1, $\overline{AD}$ is an angle bisector of $\triangle ABC$.

$$\frac{CD}{DB} = \frac{CA}{AB}$$

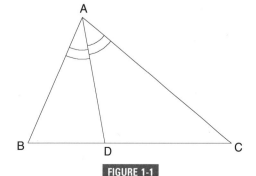

INTERACTIVE 1-1

You will be able to change the size of the triangle by grabbing vertex *A, B,* or *C* and see that the ratio is constant.

FIGURE 1-1

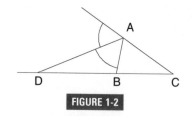

FIGURE 1-2

INTERACTIVE 1-2

You will be able to change the size of the triangle by grabbing vertex *A, B,* or *C* and see that the ratio is constant.

2. An exterior angle bisector of a triangle determines, with each of the other vertices, segments along the line containing the opposite side of the triangle that are proportional to the two remaining sides. In Figure 1-2, $\overline{AD}$ is an exterior angle bisector of $\triangle ABC$.

$$\frac{CD}{DB} = \frac{CA}{AB}$$

C. Methods of proving *triangles similar*
  1. If two triangles are similar to the same triangle, or to similar triangles, then the triangles are similar to each other.
  2. If two pairs of corresponding angles of two triangles are congruent, then the triangles are similar.
  3. If two sides of one triangle are proportional to two sides of another triangle and the angles included by those sides are congruent, then the triangles are similar.
  4. If the corresponding sides of two triangles are proportional, then the two triangles are similar.

D. *Mean proportionals* in a right triangle
  1. The altitude to the hypotenuse of a right triangle divides the hypotenuse so that either leg is the mean proportional between the hypotenuse and the segment of the hypotenuse adjacent to that leg.
  2. The altitude to the hypotenuse of a right triangle is the mean proportional between the segments of the hypotenuse.

IV. Pythagorean Theorem
A. The sum of the squares of the lengths of the legs of a right triangle equals the square of the length of the hypotenuse.

B. *Converse of the Pythagorean theorem*: If the sum of the squares of the lengths of two sides of a triangle equals the square of the length of the third side, then the angle opposite this third side is a right angle.

C. In an *isosceles right triangle*:
  1. The hypotenuse is $\sqrt{2}$ times as long as a leg.
  2. Either leg is $\dfrac{\sqrt{2}}{2}$ times as long as the hypotenuse.

D. In a *30-60-90 triangle*:
  1. The side opposite the angle of measure 30° is half as long as the hypotenuse.
  2. The side opposite the angle of measure 60° is $\dfrac{\sqrt{3}}{2}$ times as long as the hypotenuse.
  3. The hypotenuse is $\dfrac{2\sqrt{3}}{3}$ times as long as the side opposite the angle of measure 60°.
  4. The longer leg is $\sqrt{3}$ times as long as the shorter leg.

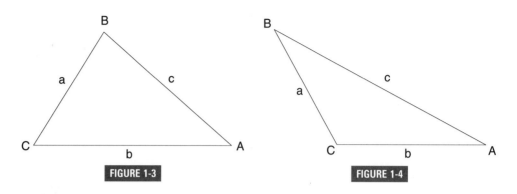

FIGURE 1-3   FIGURE 1-4

INTERACTIVE 1-3

You will be able to change the size of the triangle by grabbing vertex *A, B,* or *C* and see that the Pythagorean inequality is maintained.

E. Pythagorean inequalities
   1. In an *acute triangle* (Figure 1-3), the square of the length of any side is less than the sum of the squares of the lengths of the two remaining sides.

$$a^2 + b^2 > c^2$$

   2. In an *obtuse triangle* (Figure 1-4), the square of the length of the longest side is greater than the sum of the squares of the lengths of the two shorter sides.

$$a^2 + b^2 < c^2$$

INTERACTIVE 1-5

You will be able to change the size of the triangle or the polygons and see that the formula still holds.

F. Extension of the Pythagorean theorem: If similar polygons are constructed on the sides of a right triangle (with corresponding sides used for a side of the right triangle), then the area of the polygon on the hypotenuse equals the sum of the areas of the polygons on the legs (see Figure 1-5).

area I + area II = area III

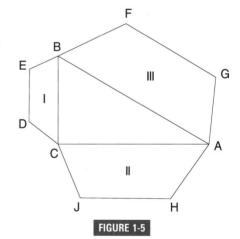

FIGURE 1-5

G. Pythagorean triples
   When $a^2 + b^2 = c^2$:

$$a = m^2 - n^2$$
$$b = 2mn$$
$$c = m^2 + n^2$$

where $m > n$.

Some common primitive Pythagorean triples are:

(3, 4, 5)	(5, 12, 13)	(7, 24, 25)	(8, 15, 17)
(9, 40, 41)	(11, 60, 61)	(12, 35, 37)	(20, 21, 29)

Note that any primitive Pythagorean triple generates an infinite number of new Pythagorean triples by all of the terms being multiplied by the same natural number.

V. Circle Relationships

A. Angle measurement with a circle

 1. The measure of an inscribed angle is one-half the measure of its intercepted arc.
 2. The measure of an angle formed by a tangent and a chord of a circle is one-half the measure of its intercepted arc.
 3. The measure of an angle formed by two chords intersecting in a point in the interior of a circle is one-half the sum of the measures of the arcs intercepted by the angle and its vertical angle.
 4. The measure of an angle formed by two secants of a circle intersecting in a point in the exterior of the circle is equal to one-half the difference of the measures of the intercepted arcs.
 5. The measure of an angle formed by a secant and a tangent to a circle intersecting in a point in the exterior of the circle is equal to one-half the difference of the measures of the intercepted arcs.
 6. The measure of an angle formed by two tangents to a circle is equal to one-half the difference of the measures of the intercepted arcs.
 7. The sum of the measure of an angle formed by two tangents to a circle and the measure of the closer intercepted arc is 180°.

An alternate way to view the seven statements above is as follows:

 1. When the vertex of an angle is a point *of* a circle, the measure of the angle is one-half the measure of the intercepted arc (see Figure 1-6).

$$m\angle APB = \frac{1}{2}x$$

INTERACTIVE 1-6

You will be able to drag points *A, B,* and *P* to change the size of the angle and see that it is still one-half the measure of the intercepted arc.

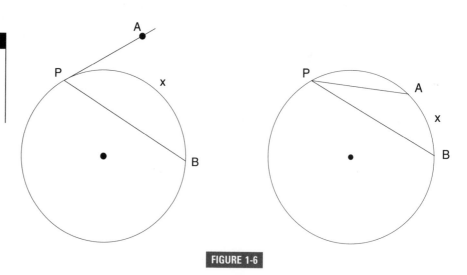

FIGURE 1-6

INTERACTIVE 1-7

You will be able to drag points *A, B, C,* and *D* to change the size of the angles and see that the formula still holds.

2. When the vertex of an angle is in the *interior* of a circle (Figure 1-7), the measure of the angle is one-half the sum of the measures of the intercepted arcs.

$$m\angle APD = \frac{1}{2}(x + y)$$

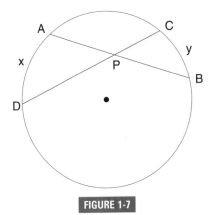

FIGURE 1-7

3. When the vertex of an angle is in the *exterior* of a circle, the measure of the angle is one-half the difference of the measures of the intercepted arcs (see Figure 1-8).

$$m\angle APB = \frac{1}{2}(x - y)$$

INTERACTIVE 1-8

You will be able to drag points *A, B, C,* and *D* to change the size of the angles and see that the formula still holds.

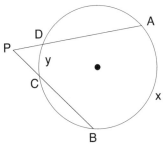
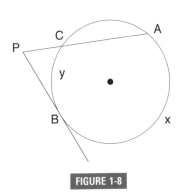
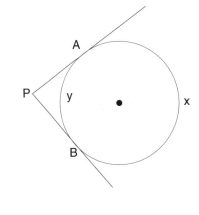

FIGURE 1-8

B. Methods of proving four points *concyclic* (a cyclic quadrilateral is a quadrilateral whose vertices are concyclic, that is, lie on the same circle)
1. If one side of a quadrilateral subtends congruent angles at the two consecutive vertices, then the quadrilateral is cyclic. Quadrilateral *ABCD* in Figure 1-9 is cyclic because $\angle DAC \cong \angle CBD$.
2. If a pair of opposite angles of a quadrilateral are supplementary, then the quadrilateral is cyclic.

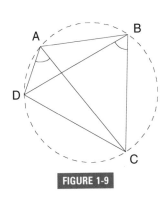

FIGURE 1-9

INTERACTIVE 1-10

You will be able to drag points *A*, *B*, and *C* to change the position of the tangent and secant and see that the formula still holds.

C. Tangent, secant, and chord segments

1. Two tangent segments that have the same endpoint in the exterior of the circle to which they are tangent are congruent.

2. If a secant segment and a tangent segment to the same circle share an endpoint in the exterior of the circle, then the square of the length of the tangent segment equals the product of the lengths of the secant segment and its external segment (see Figure 1-10).

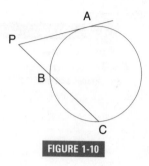

FIGURE 1-10

$$(AP)^2 = (PC)(PB)$$

INTERACTIVE 1-11

You will be able to drag points *A*, *B*, and *C* to change the position of the secants and see that the formula still holds.

3. If two secant segments of the same circle share an endpoint in the exterior of the circle, then the product of the lengths of one secant segment and its external segment equals the product of the lengths of the other secant segment and its external segment (see Figure 1-11).

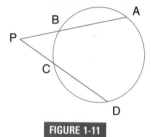

FIGURE 1-11

$$(AP)(BP) = (DP)(CP)$$

INTERACTIVE 1-12

You will be able to drag points *A*, *B, C*, and *D* to change the position of the chords and see that the formula still holds.

4. If two chords intersect in the interior of a circle, thus determining two segments in each chord, the product of the lengths of the segments of one chord equals the product of the lengths of the segments of the other chord (see Figure 1-12).

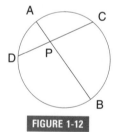

FIGURE 1-12

$$(AP)(BP) = (DP)(CP)$$

VI. Concurrency

A. The perpendicular bisectors of the sides of a triangle are concurrent at a point that is the *center of the circumscribed circle.*

B. The lines containing the three altitudes of a triangle are concurrent at a point called the *orthocenter* of the triangle.

C. The medians of a triangle are concurrent at a point of each median located two-thirds of the way from the vertex to the opposite side. This point is called the *centroid* of the triangle and is the center of gravity of the triangle.

D. The angle bisectors of a triangle are concurrent at a point that is the *center of the inscribed circle.*

VII. Inequalities

INTERACTIVE 1-13

You will be able to drag points *A*, *B*, and *C* to change the size of the triangle and see that the formula still holds.

A. The measure of an exterior angle of a triangle is greater than the measure of either remote interior angle. For $\triangle ABC$ in Figure 1-13:

$$m\angle ACD > m\angle A$$
$$m\angle ACD > m\angle B$$

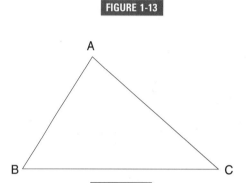

FIGURE 1-13

INTERACTIVE 1-14

You will be able to drag points *A*, *B*, and *C* to change the size of the triangle and see that the inequality still holds.

B. If two sides of a triangle are not congruent, then the angles opposite those sides are not congruent, the angle with greater measure being opposite the longer side. For $\triangle ABC$ (Figure 1-14):

If $AC > AB$, then $m\angle B > m\angle C$.

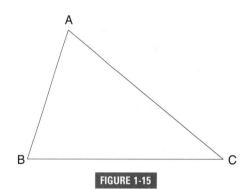

FIGURE 1-14

INTERACTIVE 1-15

You will be able to drag points *A*, *B*, and *C* to change the size of the triangle and see that the relationships still hold.

C. If two angles of a triangle are not congruent, then the sides opposite those angles are not congruent, the longer side being opposite the angle with greater measure. For $\triangle ABC$ (Figure 1-15):

If $m\angle A > m\angle C$, then $BC > AB$.

FIGURE 1-15

INTERACTIVE 1-16

You will be able to drag points *A*, *B*, and *C* to change the size of the triangle and see that the relationships still hold.

D. The sum of the lengths of any two sides of a triangle is greater than the length of the third side. For $\triangle ABC$ (Figure 1-16):

$$AB + AC > BC$$
$$AB + BC > AC$$
$$AC + BC > AB$$

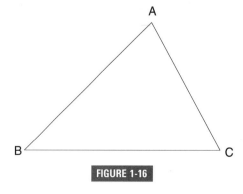 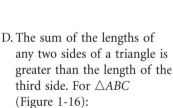

FIGURE 1-16

INTERACTIVE 1-17

You will be able to drag points *A, B, C, D, E,* and *F* to change the size of the triangles and see that the inequality still holds.

E. If two sides of a triangle are congruent respectively to two sides of a second triangle and the measure of the included angle of the first triangle is greater than the measure of the included angle of the second triangle (see Figure 1-17), then the measure of the third side of the first triangle is greater than the measure of the third side of the second triangle.

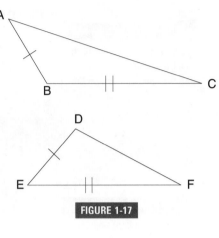

FIGURE 1-17

If $AB = DE$ and $BC = EF$ and $m\angle B > m\angle E$, then $AC > DF$.

INTERACTIVE 1-18

You will be able to drag points *A, B, C, D, E,* and *F* to change the size of the triangles and see that the inequality still holds.

F. If two sides of one triangle are congruent respectively to two sides of a second triangle and the measure of the third side of the first triangle is greater than the measure of the third side of the second triangle (see Figure 1-18), then the measure of the included angle of the first triangle is greater than the measure of the included angle of the second triangle.

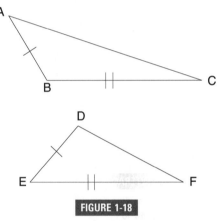

FIGURE 1-18

If $AB = DE$ and $BC = EF$ and $AC > DF$, then $m\angle B > m\angle E$.

G. In an *acute triangle*, the square of the length of any side is less than the sum of the squares of the lengths of the two remaining sides.

H. In an *obtuse triangle*, the square of the length of the longest side is greater than the sum of the squares of the lengths of the two shorter sides.

VIII. Area

A. The area of a square equals the square of the length of a side.

$$\text{area of square} = s^2$$

B. The area of a square equals one-half the square of the length of one of its diagonals.

$$\text{area of square} = \frac{1}{2}d^2$$

C. The area of any right triangle equals one-half the product of the lengths of its legs.

$$\text{area of right triangle} = \frac{1}{2}(\ell_1 \cdot \ell_2)$$

D. If two triangles have congruent bases, then the ratio of their areas equals the ratio of the lengths of their altitudes.

E. If two triangles have congruent altitudes, then the ratio of their areas equals the ratio of the lengths of their bases.

F. The area of any triangle equals one-half the product of the lengths of any two sides multiplied by the sine of the included angle.

$$\text{area of triangle} = \left(\frac{1}{2}\right)(ab) \cdot \sin \angle C$$

G. The ratio of the areas of two triangles that have one pair of congruent corresponding angles (see Figure 1-19) equals the ratio of the products of the lengths of the pairs of sides that include the angles.

For $\triangle ABC$ and $\triangle DEF$, $\angle B \cong \angle E$.

Therefore $\dfrac{\text{area } \triangle ABC}{\text{area } \triangle DEF} = \dfrac{(AB)(BC)}{(DE)(EF)}$.

INTERACTIVE 1-19

You will be able to drag points *A, B, C, D, E,* and *F* to change the size of the triangles and see that the ratios remain constant.

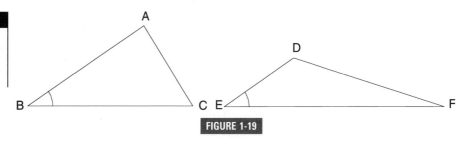

FIGURE 1-19

H. The area of an equilateral triangle equals $\dfrac{\sqrt{3}}{4}$ times the square of the length of a side.

$$\text{area of equilateral triangle} = \frac{s^2\sqrt{3}}{4}$$

I. The area of an equilateral triangle equals $\dfrac{\sqrt{3}}{3}$ times the square of the length of an altitude.

$$\text{area of equilateral triangle} = \frac{h^2\sqrt{3}}{3}$$

J. The area of any triangle with sides of length a, b, and c is $\sqrt{s(s-a)(s-b)(s-c)}$, where $s = \dfrac{a+b+c}{2}$ (s denotes the semiperimeter).

K. The area of a parallelogram equals the product of the lengths of a base and the altitude to that base.

$$\text{area of parallelogram} = b \cdot h$$

L. The area of a rhombus equals one-half the product of the lengths of its diagonals.

$$\text{area of rhombus} = \frac{1}{2}(d_1 \cdot d_2)$$

M. The area of a trapezoid equals one-half the product of the length of the altitude and the sum of the lengths of the bases.

$$\text{area of trapezoid} = \frac{1}{2}h(b_1 + b_2) = h(\text{median})$$

N. The area of a regular polygon equals one-half the product of the lengths of the apothem and the perimeter.

$$\text{area of regular polygon} = \frac{1}{2}a \cdot p$$

O. The area of a sector with radius r and a central angle of measure n equals $\dfrac{n}{360} \cdot \pi r^2$.

P. The ratio of the areas of two similar triangles equals the square of their ratio of similitude.

Q. The ratio of similitude of any pair of similar triangles equals the square root of the ratio of their areas.

R. The ratio of the areas of two similar polygons equals the square of their ratio of similitude.

S. The ratio of similitude of any pair of similar polygons equals the square root of the ratio of their areas.

Note: The ratio of similitude can be found by taking the ratio of any pair of corresponding sides, altitudes, medians, angle bisectors, or any other line segments.

LEARNING FROM GEOMETRIC FALLACIES

George Pólya, one of the great mathematicians of our time, said, *"Geometry is the science of correct reasoning on incorrect figures."* We will demonstrate in this section that making conclusions based on "incorrect" figures can lead us to impossible results. Even the statements of the fallacies sound absurd. Nevertheless, follow the "proof" of each statement and see if you can detect the mistake.

Fallacy 1, one of the more popular fallacies in Euclidean geometry, is based on the lack of a particular concept in Euclid's *Elements*.

■ FALLACY 1 Any scalene triangle is isosceles.

To prove that scalene triangle ABC is isosceles, draw the bisector of $\angle C$ and the perpendicular bisector of $\overline{AB}$. From their point of intersection, G, draw perpendiculars to $\overleftrightarrow{AC}$ and $\overleftrightarrow{CB}$, meeting these sides at points D and F, respectively.

It should be noted that there are four possibilities for the above description for various scalene triangles:

Figure 1-20, where $\overline{CG}$ and $\overline{GE}$ meet inside the triangle;

Figure 1-21, where $\overline{CG}$ and $\overline{GE}$ meet on $\overline{AB}$;

Figure 1-22, where $\overline{CG}$ and $\overline{GE}$ meet outside the triangle but the perpendiculars $\overline{GD}$ and $\overline{GF}$ fall on $\overline{AC}$ and $\overline{CB}$;

Figure 1-23, where $\overline{CG}$ and $\overline{GE}$ meet outside the triangle but the perpendiculars $\overline{GD}$ and $\overline{GF}$ meet $\overrightarrow{CA}$ and $\overrightarrow{CB}$ outside the triangle.

The "proof" of the fallacy can be done with any of these figures. Follow the "proof" on any (or all) of the figures.

GIVEN: $\triangle ABC$ is scalene.
PROVE: $AC = BC$ (or $\triangle ABC$ is isosceles)

"Proof" Because $\angle ACG \cong \angle BCG$ and right angle $CDG \cong$ right angle CFG, $\triangle CDG \cong \triangle CFG$ (SAA). Therefore $DG = FG$ and $CD = CF$. Because $AG = BG$ (a point on the perpendicular bisector of a line segment is equidistant from the endpoints of the line segment) and $\angle ADG$ and $\angle BFG$ are right angles,

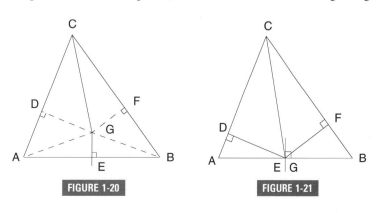

FIGURE 1-20   FIGURE 1-21

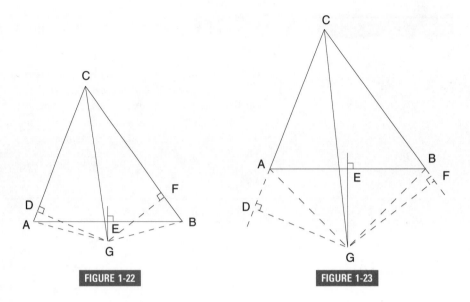

FIGURE 1-22   FIGURE 1-23

$\triangle DAG \cong \triangle FBG$ (hypotenuse–leg). Therefore $DA = FB$. It then follows that $AC = BC$ (by addition in Figures 1-20, 1-21, and 1-22, and by subtraction in Figure 1-23). ●

At this point you may be somewhat disturbed, wondering where the error was committed that permitted this fallacy to occur. By rigorous construction, you will find a subtle error in the figures:

a. The point G *must* be outside the triangle.
b. When perpendiculars meet the sides of the triangle, one will meet a side *between* the vertices, while the other will not.

INTERACTIVE 1-24

Drag points *A*, *B*, and *C* to change the shape of the triangle. Note that either *D* or *F*, but not both, always lies outside the triangle.

In general terms used by Euclid, this dilemma would remain an enigma because the concept of *betweenness* was not defined in his *Elements*. In the following discussion, we will prove that errors exist in the fallacious proof on page 13. Our proof uses Euclidean methods but assumes a definition of betweenness.

Begin by considering the circumcircle of $\triangle ABC$ (see Figure 1-24). The bisector of $\angle ACB$ must contain the midpoint G of $\overset{\frown}{AB}$ (because $\angle ACG$ and $\angle BCG$ are congruent inscribed angles). The perpendicular bisector of $\overline{AB}$ must bisect $\overset{\frown}{AB}$ and therefore must pass

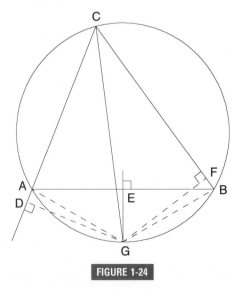

FIGURE 1-24

through point *G*. Thus the bisector of ∠*ACB* and the perpendicular bisector of $\overline{AB}$ intersect *outside* the triangle at point *G*. This eliminates the possibilities illustrated in Figures 1-20 and 1-21.

Now consider inscribed quadrilateral *ACBG*. Because the opposite angles of an inscribed (or cyclic) quadrilateral are supplementary, $m\angle CAG + m\angle CBG = 180°$. If ∠*CAG* and ∠*CBG* were right angles, then $\overline{CG}$ would be a diameter and △*ABC* would be isosceles. Therefore, because △*ABC* is scalene, ∠*CAG* and ∠*CBG* are not right angles. In this case one angle must be acute and the other obtuse. Suppose ∠*CBG* is acute and ∠*CAG* is obtuse. Then in △*CBG* the altitude on $\overline{CB}$ must be *inside* the triangle, while in obtuse triangle *CAG* the altitude on $\overline{AC}$ must be *outside* the triangle. (This is usually readily accepted without proof but can be easily proved.) The fact that one and only one of the perpendiculars intersects a side of the triangle between the vertices destroys the fallacious "proof."

❙ FALLACY 2   Two distinct perpendiculars can be drawn to a given line from a given external point.

"❶roof"   To "prove" this statement, draw any two circles, O and O′, intersecting at points *P* and *N* (Figure 1-25). Draw diameters $\overline{PA}$ and $\overline{PB}$. Then draw $\overline{AB}$ intersecting circle O at point *D* and intersecting circle O′ at point C. ∠*PDA* and ∠*PCB* are right angles because they are inscribed in semicircles of circles O and O′, respectively. Thus $\overline{PC}$ and $\overline{PD}$ are each perpendicular to $\overline{AB}$. Having two distinct lines perpendicular to a third line implies that the sum of the measures of the angles of a triangle (in this case △*PCD*) must be greater than 180°—quite disturbing in Euclidean geometry! ●

The fallacy here is created by the improper intersections of $\overline{AB}$ and the two circles. We can easily prove that the intersection of $\overline{AB}$ and the two circles is in fact at point *N*.

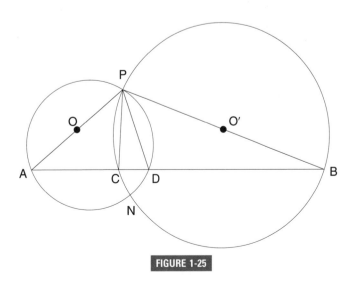

FIGURE 1-25

To do this we draw $\overline{AN}$, $\overline{BN}$, and $\overline{PN}$ (see Figure 1-26). Because $\angle ANP$ and $\angle BNP$ are inscribed in semicircles, they are both right angles. Euclid's fifth postulate ensures us that there exists a unique perpendicular to a given line through a given point on the line. Therefore the perpendiculars to $\overline{PN}$ at point N, $\overline{AN}$ and $\overline{BN}$, are simply segments of the same line, $\overleftrightarrow{ANB}$. This proves that when $\overleftrightarrow{AB}$ was first drawn it should have intersected the circles *not* in two points, C and D, but rather at one point, N, the point of intersection of the circles. Without the existence of points C and D, the fallacious proof could not have been produced.*

INTERACTIVE 1-26

Drag point P and centers O and O' to change the position of the circles.

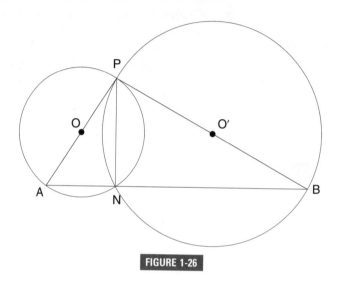

FIGURE 1-26

❚ FALLACY 3   A right angle has the same measure as an obtuse angle.

"Proof"   We begin our "proof" by drawing a rectangle $ABCD$. We then draw $\overline{CE}$ outside the rectangle so that $\overline{AD} \cong \overline{CE}$. Point P is the intersection of the perpendicular bisectors of $\overline{AE}$ and $\overline{CD}$, which intersect $\overline{AE}$ and $\overline{CD}$ at points M and N, respectively. Drawing $\overline{DP}$, $\overline{AP}$, $\overline{EP}$, and $\overline{CP}$ completes the diagram for this "proof" (Figure 1-27).

Because $AP = EP$ and $DP = CP$ (every point on the perpendicular bisector of a line segment is

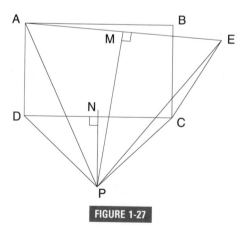

FIGURE 1-27

---

* Those who are disturbed about disproving the fallacy with the very same postulate that demonstrated the fallacy (i.e., that two *distinct* lines cannot be perpendicular to a third line) may wish to use Playfair's postulate to show that $\overline{AN}$ and $\overline{BN}$ are each parallel to $\overline{OO'}$ and hence must, in fact, be part of the same line, $\overleftrightarrow{ANB}$.

equidistant from the endpoints of the line segment), we have $\triangle ECP \cong \triangle ADP$ (SSS) and $m\angle ECP = m\angle ADP$.

However, because $\triangle PDC$ is isosceles, $m\angle DCP = m\angle CDP$. By subtraction, obtuse $\angle ECD$ has the same measure as $\angle ADC$, a right angle! ●

You may wish to consider the case when P is on $\overline{DC}$ or when P is *in* rectangle $ABCD$. Similar arguments hold for these cases.

By now you may find that an accurate construction is the best way to isolate the error in the "proof." Rather than attempt to discover the error by construction, we will analyze the situation that now exists. We notice that $\overleftrightarrow{NP}$ is also the perpendicular bisector of $\overline{AB}$. Consider $\triangle ABE$. Because $\overleftrightarrow{NP}$ and $\overleftrightarrow{MP}$ are the perpendicular bisectors of $\overline{AB}$ and $\overline{AE}$, respectively, they intersect at the center, P, of the circumcircle of $\triangle ABE$. Therefore, point P must also be on the perpendicular bisector of $\overline{BE}$.

By construction, we have $BC = EC$. Therefore point C must also lie on the perpendicular bisector of $\overline{BE}$ (see Figure 1-28). $\overleftrightarrow{PC}$ is the perpendicular bisector of $\overline{BE}$ as well as the interior angle bisector of $\angle BCE$. A reflex angle is an angle of measure greater than $180°$ and less than $360°$. Consider reflex angle ECP, whose measure is $m\angle PCR + m\angle RCE$. Thus $\overline{EP}$ in $\triangle ECP$ is placed so that it is on the side of point C *outside* the rectangle. This makes the last step of our "proof" incorrect because $m\angle ECP \neq m\angle ECD + m\angle DCP$.

INTERACTIVE 1-28

Drag points B to change the shape of the rectangle. Drag point E to reposition it.

We must keep in mind that Euclid did not use the terms *inside* and *outside* in generalized reasoning. He used these words only in reference to specific diagrams. We are able to discuss the fallacy by using these terms in general.

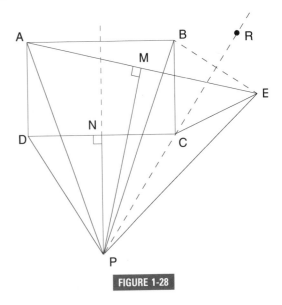

FIGURE 1-28

FALLACY 4   Every point *inside* a circle is *on* the circle.

"Proof"   Consider circle O with point P inside the circle. Choose point R on $\overrightarrow{OP}$ so that $(OP)(OR) = r^2$, where r is the length of the radius of circle O. Let the perpendicular bisector of $\overline{PR}$ intersect circle O at points S and T; let M be the midpoint of $\overline{PR}$ (Figure 1-29).

$$OP = OM - MP \tag{I}$$
$$OR = OM + MR = OM + MP \tag{II}$$

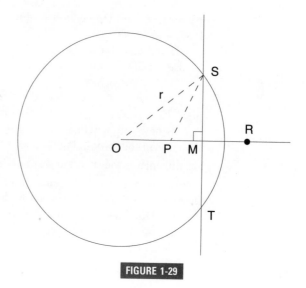

FIGURE 1-29

By multiplying (I) and (II):

$$(OP)(OR) = (OM - MP)(OM + MP)$$
$$(OP)(OR) = (OM)^2 - (MP)^2 \qquad \text{(III)}$$

By the Pythagorean theorem:

$$(OM)^2 + (MS)^2 = (OS)^2$$

or:

$$(OM)^2 = (OS)^2 - (MS)^2 \qquad \text{(IV)}$$

Also:

$$(MP)^2 + (MS)^2 = (PS)^2$$

or:

$$(MP)^2 = (PS)^2 - (MS)^2 \qquad \text{(V)}$$

Now substitute (IV) and (V) into (III) to get:

$$(OP)(OR) = [(OS)^2 - (MS)^2] - [(PS)^2 - (MS)^2]$$
$$(OP)(OR) = (OS)^2 - (PS)^2 \qquad \text{(VI)}$$

Because $\overline{OS}$ is the radius of circle O:

$$(OS)^2 = r^2 = (OP)(OR) \qquad \text{(VII)}$$

Now substitute (VII) into (VI):

$$(OP)(OR) = (OP)(OR) - (PS)^2$$

Therefore $PS = 0$, which implies that point P must be *on* the circle! ●

To discover the fallacy in this "proof," we let $OP = a$. Therefore $OR = \dfrac{r^2}{a}$.
Because $r > a$ and the square of a real number is positive, $(r - a)^2 > 0$. This
can be written as $r^2 - 2ra + a^2 > 0$. Thus $r^2 + a^2 > 2ra$. Multiplying both sides
of this inequality by $\dfrac{1}{2a}$, we get $\dfrac{1}{2}\left(\dfrac{r^2}{a} + a\right) > r$, which is $\dfrac{1}{2}(OR + OP) > r$, or
$OM > r$. This implies that point M must be *outside* the circle and that points S
and T do not exist. This destroys the fallacious "proof."

■ **FALLACY 5**   Two segments of unequal length are actually of equal length.

"℗roof"   Consider $\triangle ABC$, with $\overline{MN} \parallel \overline{BC}$
and $\overline{MN}$ intersecting $\overline{AB}$ and $\overline{AC}$
in points M and N, respectively
(see Figure 1-30). We will now
"prove" that $\overline{BC} = \overline{MN}$.

Because $\overline{MN} \parallel \overline{BC}$, we have
$\triangle AMN \sim \triangle ABC$ and $\dfrac{BC}{MN} = \dfrac{AB}{AM}$.
It then follows that $(BC)(AM) = (AB)(MN)$. Now multiply both
sides of this equality by
$(BC - MN)$ to get:

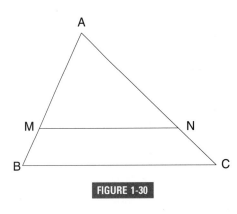

FIGURE 1-30

$$(BC)^2(AM) - (BC)(AM)(MN) = (AB)(MN)(BC) - (AB)(MN)^2$$

By adding $(BC)(AM)(MN) - (AB)(MN)(BC)$ to both sides, we get:

$$(BC)^2(AM) - (AB)(MN)(BC) = (BC)(AM)(MN) - (AB)(MN)^2$$

This equation can be written as:

$$(BC)[(BC)(AM) - (AB)(MN)] = (MN)[(BC)(AM) - (AB)(MN)]$$

By dividing both sides by the common factor $[(BC)(AM) - (AB)(MN)]$, we find
that $BC = MN$! ●

No discussion of mathematical fallacies would be complete without an example of a dilemma resulting from division by zero. We committed this mathematical sin when we divided by zero in the form of $[(BC)(AM) - (AB)(MN)]$, which
was a consequence of the triangles proved to be similar earlier.

COMMON NOMENCLATURE

Figure 1-31 illustrates some of the details we will consider in this book. We list them systematically now, with the general understanding that we may use a symbol ambiguously when we can simplify our work without confusion. Thus we may use b to represent either a side of a triangle, its name, or its measure, as the context should make clear. The ambiguity reflects our choice and not our ignorance—our aim is clarity. The rigor and precision that support the material could certainly be supplied, but only with time and space that seem inappropriate in our discussion.

Sides: a, b, c

Angles: α, β, γ

Vertices: A, B, C

Altitudes: h_a, h_b, h_c

Feet of the altitudes: H_a, H_b, H_c

Orthocenter (point of concurrence of altitudes): H

Medians: m_a, m_b, m_c

Midpoints of sides: M_a, M_b, M_c

Centroid (point of concurrence of medians): G

Angle bisectors: t_a, t_b, t_c

Feet of angle bisectors: T_a, T_b, T_c

Incenter (point of concurrence of angle bisectors; center of inscribed circle): I

Inradius (radius of inscribed circle): r

Circumcenter (point of concurrence of perpendicular bisectors of sides; center of circumscribed circle): O

Circumradius (radius of circumscribed circle): R

Semiperimeter (half the sum of the lengths of the sides: $\frac{1}{2}(a + b + c)$): s

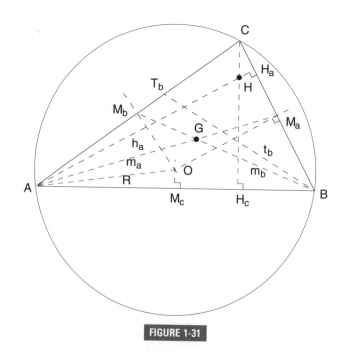

FIGURE 1-31

── EXERCISES ──

1. Discover the fallacy in the following "proof": *If two opposite sides of a quadri-lateral are congruent, then the remaining two sides must be parallel.*

"Proof"   In quadrilateral $ABCD$, $AD = BC$. Construct perpendicular bisectors $\overline{OP}$ and $\overline{OQ}$ of sides $\overline{DC}$ and $\overline{AB}$ at points P and Q, respectively. Point N is on $\overrightarrow{PO}$. (We let O be the intersection of the two perpendicular bisectors of given nonparallels, $\overline{OP}$ and $\overline{OQ}$.) (See Figure 1-32.) Because O is a point on the perpendicular bisector of $\overline{DC}$, $\overline{DO} \cong \overline{CO}$. Similarly, $\overline{OA} \cong \overline{OB}$. We began with $AD = BC$. Therefore $\triangle ADO \cong \triangle BCO$ (SSS) and $m\angle AOD = m\angle BOC$.

We can easily establish that $m\angle DOP = m\angle COP$. By addition, $m\angle AOP = m\angle BOP$. The supplements of these angles are also equal in measure: $m\angle AON = m\angle BON$. But because $\triangle AOQ \cong \triangle BOQ$ (SSS), $m\angle AOQ = m\angle BOQ$. Because the angle bisector is unique, $\overrightarrow{ON}$ and $\overrightarrow{OQ}$ must coincide and the perpendiculars to these must also be parallel. Hence $\overline{AB} \parallel \overline{CD}$. ●

Repeat the "proof" for O *outside* the quadrilateral. Then repeat the proof for O on $\overline{DC}$.

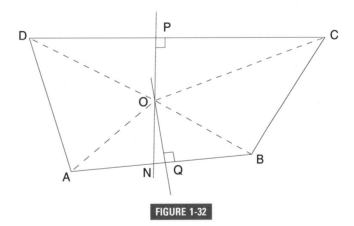

FIGURE 1-32

2. Discover the fallacy in the following "proof": $45° = 60°$.

"Proof" Construct equilateral triangle ABC. (See Figure 1-33). On side $\overline{AB}$ construct isosceles right triangle ADB with $\overline{AB}$ as hypotenuse. Lay off $\overline{EB}$ on $\overline{BC}$ equal in length to $\overline{BD}$. Connect point E to point F, the midpoint of $\overline{AD}$, and extend to meet $\overrightarrow{AB}$ at point G. Draw $\overline{GD}$. Construct perpendicular bisectors of $\overline{GD}$ and $\overline{GE}$. Because $\overline{GD}$ and $\overline{GE}$ are not parallel, the perpendicular bisectors must meet at point K. Connect point K with points G, D, E, and B.

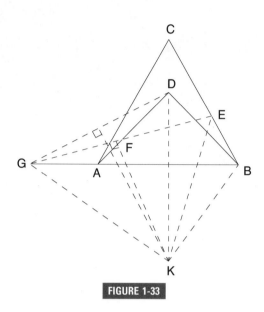

FIGURE 1-33

Because $GK = KD$ and $GK = KE$ (a point on the perpendicular bisector of a line segment is equidistant from the ends of a line segment), $KD = KE$. We constructed $DB = EB$. Therefore $\triangle KBD \cong \triangle KBE$ (SSS) and $m\angle KBD = m\angle KBE$. By subtraction, $m\angle DBG = m\angle EBG$. But $m\angle DBG = 45°$, while $m\angle CBG = 60°$; thus $45° = 60°$. ●

3. Parallelograms $ABGF$ and $ACDE$ are constructed on sides $\overline{AB}$ and $\overline{AC}$ of $\triangle ABC$ (see Figure 1-34). ($\triangle ABC$ is any type of triangle.) $\overrightarrow{DE}$ and $\overrightarrow{GF}$ intersect at point P. Using $\overline{BC}$ as a side, construct parallelogram $BCJK$ so that $\overline{BK} \parallel \overline{PA}$ and $\overline{BK} \cong \overline{PA}$. From this configuration, Pappus (A.D. 300) proposed an extension of the Pythagorean theorem. He proved that the sum of the area of parallelogram $ABGF$ and the area of parallelogram $ACDE$ is equal to the area of parallelogram $BCJK$. Prove this relationship. (Note: You may wish to model your proof after Euclid's proof of the Pythagorean theorem.)

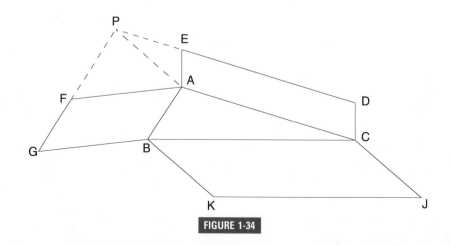

FIGURE 1-34

4. *GIVEN*: $\overline{BE}$ and $\overline{AD}$ are altitudes (intersecting at H) of $\triangle ABC$, while F, G, and K are midpoints of $\overline{AH}$, $\overline{AB}$, and $\overline{BC}$, respectively (see Figure 1-35).

   PROVE: $\angle FGK$ is a right angle.

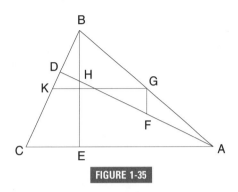

FIGURE 1-35

5. A line $\overleftrightarrow{PQ}$, parallel to base $\overline{BC}$ of $\triangle ABC$, intersects $\overline{AB}$ and $\overline{AC}$ at points P and Q, respectively (see Figure 1-36). The circle passing through P and tangent to $\overline{AC}$ at Q intersects $\overline{AB}$ again at point R. Prove that points R, Q, C, and B are concyclic.

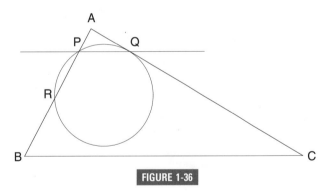

FIGURE 1-36

   As you proceed through the rest of this book, you may want to work with additional exercises. For this purpose you might use *Challenging Problems in Geometry* by A. S. Posamentier and C. T. Salkind (New York: Dover, 1996).

CONCURRENCY OF LINES IN A TRIANGLE

INTRODUCTION

In spite of its importance, the concept of concurrency of lines (i.e., three or more lines containing a common point) usually gets a light treatment in an elementary geometry course because of higher priorities. Acquiring a truly good facility with the concept would require that more theorems be explored than time permits in the first geometry course. Familiar concurrencies such as the medians, angle bisectors, and altitudes of a triangle are mentioned but not often established by proof. Introducing a few new theorems makes the topic of concurrency quite simple and presents a new vista in Euclidean geometry. This chapter begins by demonstrating the importance of establishing concurrency. With the help of an important theorem, first published by Giovanni Ceva in 1678, we present a variety of interesting relationships and theorems. You will soon see how simply some previously difficult theorems can be proved.

Because we want to show the importance of concurrency, let us consider the following problem:

Two wires are placed in straight lines meeting in an inaccessible region (Figure 2-1). How would you locate the proper placement for a wire that is to *bisect* the angle formed by the two wires without touching the inaccessible region?

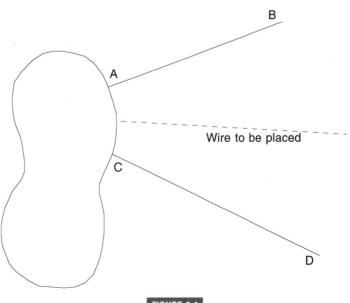

FIGURE 2-1

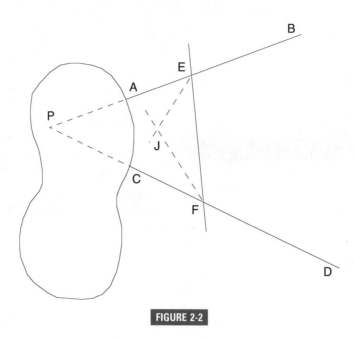

FIGURE 2-2

Although there are many possible methods of solution to this problem, we chose the following solution for a reason that will soon become clear.

Solution Draw any line through $\overleftrightarrow{AB}$ and $\overleftrightarrow{CD}$, intersecting them at points E and F, respectively (see Figure 2-2). Construct the bisectors of $\angle AEF$ and $\angle CFE$, which meet at point J. Suppose $\triangle PEF$ were complete. The bisector of $\angle P$ would then have to contain point J because the angle bisectors of a triangle are concurrent.

Repeat this process for any other line $\overleftrightarrow{GH}$ that meets $\overleftrightarrow{AB}$ and $\overleftrightarrow{CD}$ at points G and H, respectively (Figure 2-3). This time, bisect $\angle AGH$ and $\angle CHG$. These bisectors meet at point K. Once again, we notice that the required angle bisector (that of $\angle P$) must contain point K. Because this required angle bisector *must* contain both J and K, these two points determine our desired line, which is the location of the wire to be installed. ●

This solution relies heavily on the notion that the angle bisectors of a triangle are concurrent. As we have said, the topic of concurrency in a triangle deserves more attention than it usually gets in the elementary geometry course. In a very simple way, we will prove that the angle bisectors of a triangle are concurrent. First, we must establish an extremely useful relationship.

Recall from elementary geometry that there are many "centers" of a triangle. Some examples are:

- *centroid*—the center of gravity of the triangle, determined by the intersection of the medians;
- *orthocenter*—the point of intersection of the altitudes of the triangle;
- *incenter*—the center of the inscribed circle of the triangle, determined by the intersection of the angle bisectors of the triangle;

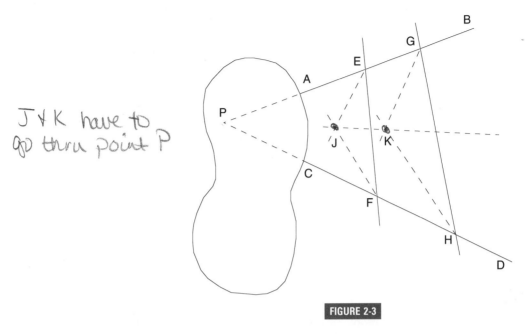

J & K have to go thru point P

FIGURE 2-3

- *circumcenter*—the center of the circumscribed circle (or circumcircle), determined by the intersection of the perpendicular bisectors of the sides of the triangle.

Numerous applications of these triangle centers are considered in the elementary geometry course. On occasion, students will consider "practical" applications that rely on the concurrency property of these points. We offered one such application at the beginning of this chapter. Yet because the traditional proofs of these concurrency relationships are somewhat cumbersome, the relationships are frequently accepted without proof. With the help of the famous theorem first published* by the Italian mathematician Giovanni Ceva (1647–1734), which bears his name, we will produce simple proofs of the concurrency relationships named previously, as well as many others.

CEVA'S THEOREM

Know how to form product

THEOREM 2.1   (**Ceva's Theorem**) The three lines containing the vertices *A*, *B*, and *C* of △*ABC* and intersecting the opposite sides at points *L*, *M*, and *N*, respectively, are concurrent if and only if $\dfrac{AN}{NB} \cdot \dfrac{BL}{LC} \cdot \dfrac{CM}{MA} = 1$.

---

* *De lineis se invicem secantibus statica constructio* (Milan, 1678).

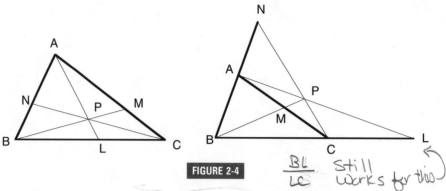

FIGURE 2-4

$\frac{BL}{LC}$ still works for this

To prove this theorem, we must note that there are two possible situations in which the three lines drawn from the vertices may intersect the sides and still be concurrent. These situations are pictured in Figure 2-4. It is perhaps easier to follow the proof with the diagram on the left and verify the validity of the statements with the diagram on the right. In any case, the statements made in the proof hold for *both* diagrams.

Ceva's theorem is an equivalence (or biconditional) and therefore requires two proofs (one the converse of the other). We will first prove that *if the three lines containing the vertices of* $\triangle ABC$ *and intersecting the opposite sides at points L, M, and N, respectively, are concurrent, then* $\frac{AN}{NB} \cdot \frac{BL}{LC} \cdot \frac{CM}{MA} = 1$. We offer three proofs. The first (though not the simplest) requires no auxiliary lines.

Proof I   In Figure 2-4, $\overline{AL}$, $\overline{BM}$, and $\overline{CN}$ meet at point P. Because $\triangle ABL$ and $\triangle ACL$ share the same altitude (i.e., from point A):

$$\frac{\text{area } \triangle ABL}{\text{area } \triangle ACL} = \frac{BL}{LC} \tag{I}$$

Similarly:

$$\frac{\text{area } \triangle PBL}{\text{area } \triangle PCL} = \frac{BL}{LC} \tag{II}$$

From (I) and (II):

$$\frac{\text{area } \triangle ABL}{\text{area } \triangle ACL} = \frac{\text{area } \triangle PBL}{\text{area } \triangle PCL}$$

A basic property of proportions $\left(\frac{w}{x} = \frac{y}{z} = \frac{w-y}{x-z}\right)$ provides that:

$$\frac{BL}{LC} = \frac{\text{area } \triangle ABL - \text{area } \triangle PBL}{\text{area } \triangle ACL - \text{area } \triangle PCL} = \frac{\text{area } \triangle ABP}{\text{area } \triangle ACP} \tag{III}$$

We now repeat the process, using $\overline{BM}$ instead of $\overline{AL}$:

$$\frac{CM}{MA} = \frac{\text{area } \triangle BMC}{\text{area } \triangle BMA} = \frac{\text{area } \triangle PMC}{\text{area } \triangle PMA}$$

It follows that:

$$\frac{CM}{MA} = \frac{\text{area } \triangle BMC - \text{area } \triangle PMC}{\text{area } \triangle BMA - \text{area } \triangle PMA} = \frac{\text{area } \triangle BCP}{\text{area } \triangle BAP} \qquad (IV)$$

Once again we repeat the process, this time using $\overline{CN}$ instead of $\overline{AL}$:

$$\frac{AN}{NB} = \frac{\text{area } \triangle ACN}{\text{area } \triangle BCN} = \frac{\text{area } \triangle APN}{\text{area } \triangle BPN}$$

This gives us:

$$\frac{AN}{NB} = \frac{\text{area } \triangle ACN - \text{area } \triangle APN}{\text{area } \triangle BCN - \text{area } \triangle BPN} = \frac{\text{area } \triangle ACP}{\text{area } \triangle BCP} \qquad (V)$$

We now simply multiply (III), (IV), and (V) to get the desired result:

$$\frac{BL}{LC} \cdot \frac{CM}{MA} \cdot \frac{AN}{NB} = \frac{\text{area } \triangle ABP}{\text{area } \triangle ACP} \cdot \frac{\text{area } \triangle BCP}{\text{area } \triangle BAP} \cdot \frac{\text{area } \triangle ACP}{\text{area } \triangle BCP} = 1 \ \bullet$$

By introducing an auxiliary line, we can produce a simpler proof.

⊘roof II Consider Figure 2-4, but add a line containing point A and parallel to $\overline{BC}$ that intersects $\overleftrightarrow{CP}$ at point S and $\overleftrightarrow{BP}$ at point R (see Figure 2-5). The parallel lines enable us to establish the following pairs of similar triangles:

$$\triangle AMR \sim \triangle CMB \Rightarrow \frac{AM}{MC} = \frac{AR}{CB} \qquad (I)$$

$$\triangle BNC \sim \triangle ANS \Rightarrow \frac{BN}{NA} = \frac{CB}{SA} \qquad (II)$$

$$\triangle CLP \sim \triangle SAP \Rightarrow \frac{CL}{SA} = \frac{LP}{AP} \qquad (III)$$

$$\triangle BLP \sim \triangle RAP \Rightarrow \frac{BL}{RA} = \frac{LP}{AP} \qquad (IV)$$

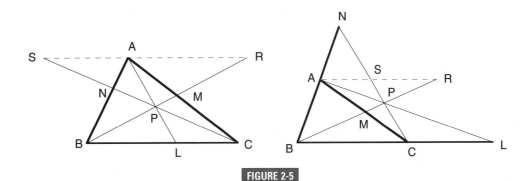

FIGURE 2-5

From (III) and (IV) we get:

$$\frac{CL}{SA} = \frac{BL}{RA}$$

This can be rewritten as:

$$\frac{CL}{BL} = \frac{SA}{RA} \qquad (V)$$

By multiplying (I), (II), and (V), we obtain our desired result:

$$\frac{AM}{MC} \cdot \frac{BN}{NA} \cdot \frac{CL}{BL} = \frac{AR}{CB} \cdot \frac{CB}{SA} \cdot \frac{SA}{RA} = 1$$

We rearrange the terms and invert the fractions to get:

$$\frac{AN}{NB} \cdot \frac{BL}{LC} \cdot \frac{CM}{MA} = 1 \text{ (the same as the conclusion of Proof I) } \bullet$$

By adding two auxiliary lines to the diagrams in Figure 2-4, we are able to produce another proof, again using the properties of similar triangles.

Proof III   We begin with the diagrams shown in Figure 2-4 but add two lines to each diagram. We draw a line through point A and a line through point C each parallel to $\overline{BP}$ and intersecting $\overleftrightarrow{CP}$ and $\overleftrightarrow{AP}$ at points S and R, respectively (see Figure 2-6).

$$\triangle ASN \sim \triangle BPN \Rightarrow \frac{AN}{NB} = \frac{AS}{BP} \qquad (I)$$

$$\triangle BPL \sim \triangle CRL \Rightarrow \frac{BL}{LC} = \frac{BP}{CR} \qquad (II)$$

$$\triangle PAM \sim \triangle RAC \Rightarrow \frac{CA}{MA} = \frac{RC}{PM}$$

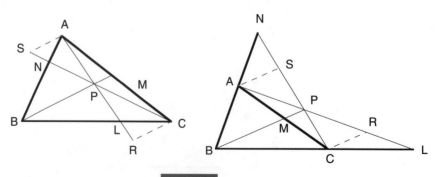

FIGURE 2-6

This can be written as:

$$CA = \frac{(RC)(MA)}{PM} \qquad \text{(III)}$$

$$\triangle PCM \sim \triangle SCA \Rightarrow \frac{CM}{CA} = \frac{PM}{AS}$$

This can be written as:

$$CA = \frac{(AS)(CM)}{PM} \qquad \text{(IV)}$$

From (III) and (IV):

$$\frac{(RC)(MA)}{PM} = \frac{(AS)(CM)}{PM}$$

This can be written as:

$$\frac{CM}{MA} = \frac{RC}{AS} \qquad \text{(V)}$$

To obtain our desired result, we multiply (I), (II), and (V):

$$\frac{AN}{NB} \cdot \frac{BL}{LC} \cdot \frac{CM}{MA} = \frac{AS}{BP} \cdot \frac{BP}{CR} \cdot \frac{RC}{AS} = 1 \; \bullet$$

To complete the proof of Ceva's theorem, we must now prove the converse of the implication proved above; that is, we will now prove that *if the lines containing the vertices of $\triangle ABC$ intersect the opposite sides in points L, M, and N, respectively, so that $\frac{AN}{NB} \cdot \frac{BL}{LC} \cdot \frac{CM}{MA} = 1$, then these lines, $\overleftrightarrow{AL}, \overleftrightarrow{BM},$ and $\overleftrightarrow{CN},$ are concurrent.*

❷**roof**   Suppose $\overleftrightarrow{BM}$ and $\overleftrightarrow{AL}$ intersect at point *P.* Draw $\overleftrightarrow{PC}$ and call its intersection with $\overleftrightarrow{AB}$ N'. Now that $\overleftrightarrow{AL}$, $\overleftrightarrow{BM}$, and $\overleftrightarrow{CN'}$ are concurrent, we can use the part of Ceva's theorem proved earlier to state the following:

$$\frac{AN'}{N'B} \cdot \frac{BL}{LC} \cdot \frac{CM}{MA} = 1$$

Our hypothesis stated that:

$$\frac{AN}{NB} \cdot \frac{BL}{LC} \cdot \frac{CM}{MA} = 1$$

Therefore $\frac{AN'}{N'B} = \frac{AN}{NB}$, so *N* and *N'* must coincide, proving concurrency. ●

APPLICATIONS OF CEVA'S THEOREM

One of the best ways to show the usefulness of Ceva's theorem is to apply it to the proof of the concurrency of the various line segments encountered in elementary geometry. The simplest application is to prove the concurrency of the medians of a triangle. To best appreciate the "power" of Ceva's theorem, you should first recall the conventional method of proving the medians of a triangle concurrent. Suffice it to say that it is quite long and complex. Compared to this rather cumbersome proof, the method we use here should provoke some excitement about Ceva's theorem.

Ⓐpplication 1   Prove that the medians of a triangle are concurrent. ●

Ⓟroof   In $\triangle ABC$, $\overline{AL}$, $\overline{BM}$, and $\overline{CN}$ are medians (see Figure 2-7). Therefore $AN = NB$, $BL = LC$, and $CM = MA$. Multiplying these equalities gives us:

$$(AN)(BL)(CM) = (NB)(LC)(MA) \quad \text{or} \quad \frac{AN}{NB} \cdot \frac{BL}{LC} \cdot \frac{CM}{MA} = 1$$

INTERACTIVE 2-7

Drag vertices *A, B*, and *C* to change the shape of the triangle and see that the medians always meet at one point.

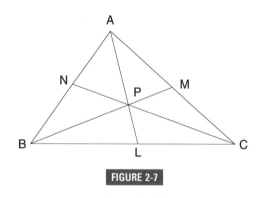

FIGURE 2-7

Thus by Ceva's theorem, $\overline{AL}$, $\overline{BM}$, and $\overline{CN}$ are concurrent. ●

Again, it would be advisable to compare the conventional proof (that presented in the context of elementary geometry) for the concurrency of the altitudes of a triangle to the following proof, which uses Ceva's theorem.

Ⓐpplication 2   Prove that the altitudes of a triangle are concurrent. ●

Ⓟroof   In $\triangle ABC$, $\overline{AL}$, $\overline{BM}$, and $\overline{CN}$ are altitudes. You can follow this proof for both diagrams of Figure 2-8 because the same proof holds true for both an acute and an obtuse triangle.

$$\triangle ANC \sim \triangle AMB \Rightarrow \frac{AN}{MA} = \frac{AC}{AB} \tag{I}$$

$$\triangle BLA \sim \triangle BNC \Rightarrow \frac{BL}{NB} = \frac{AB}{BC} \tag{II}$$

$$\triangle CMB \sim \triangle CLA \Rightarrow \frac{CM}{LC} = \frac{BC}{AC} \tag{III}$$

INTERACTIVE 2-8

Drag vertices *A, B,* and *C* to change the shape of the triangle and see that the altitudes always meet at one point.

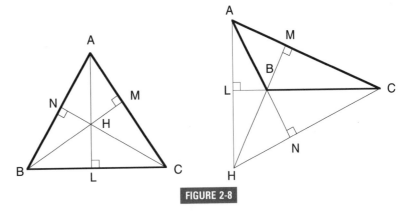

FIGURE 2-8

Multiplying (I), (II), and (III) gives us:

$$\frac{AN}{MA} \cdot \frac{BL}{NB} \cdot \frac{CM}{LC} = \frac{AC}{AB} \cdot \frac{AB}{BC} \cdot \frac{BC}{AC} = 1$$

Therefore the altitudes are concurrent (by Ceva's theorem). ●

The proof that the three angle bisectors of a triangle are concurrent is left as an exercise. The following proof should be helpful in working that exercise.

Ⓐpplication 3   Prove that the bisector of any interior angle of a nonisosceles triangle and the bisectors of the two exterior angles at the other vertices are concurrent. ●

Ⓟroof   In $\triangle ABC$, $\overleftrightarrow{AL}$ bisects $\angle BAC$ and meets $\overleftrightarrow{BC}$ at point L, $\overleftrightarrow{BM}$ bisects exterior $\angle ABE$ and meets $\overleftrightarrow{AC}$ at point M, and $\overleftrightarrow{CN}$ bisects exterior $\angle ACF$ and meets $\overleftrightarrow{AB}$ at point N (see Figure 2-9).

Because the bisector $(\overleftrightarrow{AL})$ of an interior angle of a triangle partitions the opposite side proportionally to the remaining two sides of the triangle:

$$\frac{BL}{LC} = \frac{AB}{AC} \qquad\qquad (I)$$

An exterior angle bisector partitions the side that it intersects proportionally to the remaining sides of the triangle. This property produces the following proportions:

$$\text{For } \overleftrightarrow{BM}: \quad \frac{CM}{MA} = \frac{BC}{AB} \qquad\qquad (II)$$

$$\text{For } \overleftrightarrow{CN}: \quad \frac{AN}{NB} = \frac{AC}{BC} \qquad\qquad (III)$$

By multiplying (I), (II), and (III), we get:

$$\frac{BL}{LC} \cdot \frac{CM}{MA} \cdot \frac{AN}{NB} = \frac{AB}{AC} \cdot \frac{BC}{AB} \cdot \frac{AC}{BC} = 1$$

INTERACTIVE 2-9

Drag vertices *A*, *B*, and *C* to
change the shape of the
triangle and see that the
indicated bisectors of the angles
(interior and exterior) always
meet at one point.

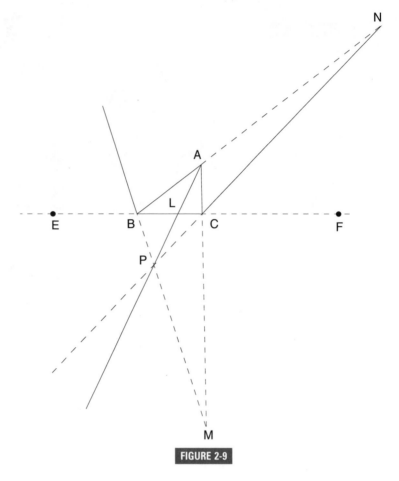

FIGURE 2-9

By Ceva's theorem, we may conclude that $\overleftrightarrow{AL}$, $\overleftrightarrow{BM}$, and $\overleftrightarrow{CN}$ are concurrent. ●

Sometimes the question of concurrency is a bit disguised, as in the following application.

Ⓐpplication 4 In $\triangle ABC$, $\overline{PQ} \parallel \overline{BC}$ and intersects $\overline{AB}$ and $\overline{AC}$ at points P and Q, respectively (see Figure 2-10). Prove that $\overline{PC}$ and $\overline{QB}$ intersect at a point on median $\overline{AM}$. ●

INTERACTIVE 2-10

Drag vertices *A*, *B,* and *C* to
change the shape of the triangle;
drag point *P* on $\overline{AB}$ and see that
$\overline{PC}$ and $\overline{QB}$ always meet at a
point on $\overline{AM}$.

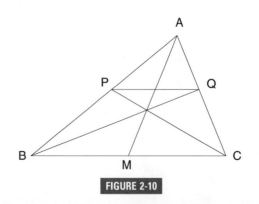

FIGURE 2-10

Proof   Because $\overline{PQ} \parallel \overline{BC}$:

$$\frac{AP}{PB} = \frac{AQ}{QC} \quad \text{or} \quad \frac{AP}{PB} \cdot \frac{QC}{AQ} = 1 \tag{I}$$

Because $\overline{AM}$ is a median, $BM = MC$. Therefore:

$$\frac{BM}{MC} = 1 \tag{II}$$

By multiplying (I) and (II), we get:

$$\frac{AP}{PB} \cdot \frac{QC}{AQ} \cdot \frac{BM}{MC} = 1$$

Thus, by Ceva's theorem, $\overline{AM}$, $\overline{QB}$, and $\overline{PC}$ are concurrent, or $\overline{QB}$ and $\overline{PC}$ intersect at a point on $\overline{AM}$. ●

Up to this point, all our applications have been used to prove concurrency. The following application demonstrates a somewhat different use of Ceva's theorem.

Application 5   In $\triangle ABC$, where $\overline{CD}$ is the altitude to $\overline{AB}$ and P is any point on $\overline{DC}$, $\overline{AP}$ intersects $\overline{CB}$ at point Q and $\overline{BP}$ intersects $\overline{CA}$ at point R (see Figure 2-11). Prove that $\angle RDC \cong \angle QDC$. ●

Proof   Let $\overrightarrow{DR}$ and $\overrightarrow{DQ}$ intersect the line containing C and parallel to $\overline{AB}$, at points G and H, respectively.

$$\triangle CGR \sim \triangle ADR \Rightarrow \frac{CR}{RA} = \frac{GC}{AD} \tag{I}$$

$$\triangle BDQ \sim \triangle CHQ \Rightarrow \frac{BQ}{QC} = \frac{DB}{CH} \tag{II}$$

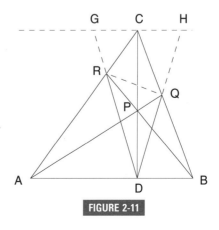

FIGURE 2-11

We now apply Ceva's theorem to $\triangle ABC$ to get:

$$\frac{CR}{RA} \cdot \frac{AD}{DB} \cdot \frac{BQ}{QC} = 1 \qquad \text{(III)}$$

Substituting (I) and (II) into (III) gives us:

$$\frac{GC}{AD} \cdot \frac{AD}{DB} \cdot \frac{DB}{CH} = 1 \quad \text{or} \quad \frac{GC}{CH} = 1$$

This implies that $GC = CH$. Thus $\overline{CD}$ is the perpendicular bisector of $\overline{GH}$. Hence $\triangle GCD \cong \triangle HCD$, and therefore $\angle RDC \cong \angle QDC$. ●

From the preceding applications, we have seen how Ceva's theorem easily enables us to prove theorems whose proofs would otherwise be quite complex. Ceva's theorem again demonstrates its usefulness in assisting us in proving an interesting point of concurrency in a triangle known as the *Gergonne point*.

THE GERGONNE POINT

A fascinating point of concurrency in a triangle was first established by French mathematician Joseph-Diaz Gergonne (1771–1859). Gergonne reserved a distinct place in the history of mathematics as the initiator (in 1810) of the first purely mathematical journal, *Annales des mathématiques pures et appliqués*. The journal appeared monthly until 1832 and was known as *Annales del Gergonne*. During the time of its publication, Gergonne published about two hundred papers, mostly on geometry. Gergonne's *Annales* played an important role in the establishment of projective and algebraic geometry by giving some of the greatest minds of the times an opportunity to share information. Here we consider a rather simple theorem established by Gergonne that exhibits concurrency and is easily proved using Ceva's theorem.

THEOREM 2.2 The lines containing a vertex of a triangle and the point of tangency of the opposite side with the inscribed circle are concurrent. (This point of concurrency is known as the **Gergonne point** of the triangle.)

Proof Circle O is tangent to sides $\overline{AB}$, $\overline{AC}$, and $\overline{BC}$ of $\triangle ABC$ at points N, M, and L, respectively (see Figure 2-12). It follows that $AN = AM$, $BL = BN$, and $CM = CL$. Each of these equalities can be written as:

$$\frac{AN}{AM} = 1 \qquad \frac{BL}{BN} = 1 \qquad \frac{CM}{CL} = 1$$

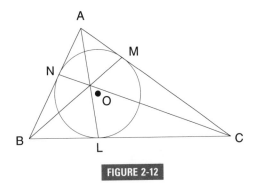

FIGURE 2-12

By multiplying these three fractions, we get:

$$\frac{AN}{AM} \cdot \frac{BL}{BN} \cdot \frac{CM}{CL} = 1$$

Therefore:

$$\frac{AN}{BN} \cdot \frac{BL}{CL} \cdot \frac{CM}{AM} = 1$$

By Ceva's theorem, this equality implies that $\overline{AL}$, $\overline{BM}$, and $\overline{CN}$ are concurrent. The point of concurrency is the Gergonne point of $\triangle ABC$. ●

E X E R C I S E S

1. Prove that the angle bisectors of a triangle are concurrent.

2. If point P is situated on $\overline{BC}$ so that $AB + BP = AC + CP$, point Q is situated on $\overline{AC}$ so that $BC + CQ = AB + AQ$, and point R is situated on $\overline{AB}$ so that $BC + BR = AC + AR$, prove that $\overline{AP}$, $\overline{BQ}$, and $\overline{CR}$ are concurrent (see Figure 2-13). (This point of concurrency is known as the *Nagel point** of $\triangle ABC$.)

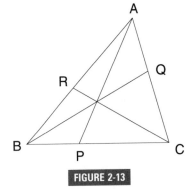

FIGURE 2-13

* Discovered by C. H. Nagel (1803–1882), the point can also be described as the intersection of the lines from the vertices of a triangle to the points of tangency of the opposite escribed circles.

3. $\triangle ABC$ cuts a circle at points E, E', D, D', F, and F', as in Figure 2-14. Prove that if $\overline{AD}$, $\overline{BF}$, and $\overline{CE}$ are concurrent, then $\overline{AD'}$, $\overline{BF'}$, and $\overline{CE'}$ are also concurrent.

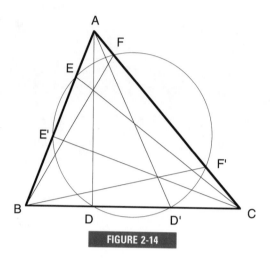

FIGURE 2-14

4. In $\triangle ABC$ (Figure 2-15), $\overline{AL}$, $\overline{BM}$, and $\overline{CN}$ are concurrent at point P. Points R, S, and T are chosen on $\overline{BC}$, $\overline{AC}$, and $\overline{AB}$, respectively, so that $\overline{NR} \parallel \overline{AC}$, $\overline{LS} \parallel \overline{AB}$, and $\overline{MT} \parallel \overline{BC}$. Prove that $\overline{AR}$, $\overline{BS}$, and $\overline{CT}$ are concurrent (at point Q).

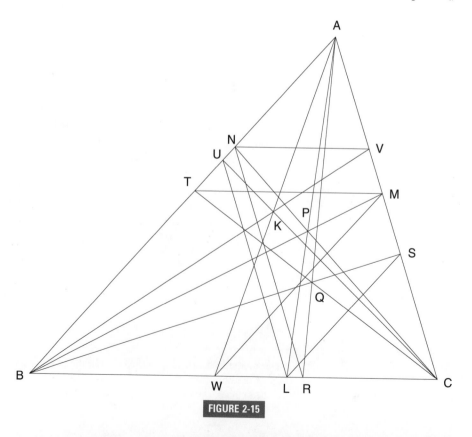

FIGURE 2-15

5. In △ABC (Figure 2-15), $\overline{AL}$, $\overline{BM}$, and $\overline{CN}$ are concurrent at point P. Points U, V, and W are chosen on $\overline{AB}$, $\overline{AC}$, and $\overline{BC}$, respectively, so that $\overline{LU} \parallel \overline{AC}$, $\overline{NV} \parallel \overline{BC}$, and $\overline{MW} \parallel \overline{AB}$. Prove that $\overline{AW}$, $\overline{BV}$, and $\overline{CU}$ are concurrent (at point K).

6. In △ABC (Figure 2-16), $\overline{AL}$, $\overline{BM}$, and $\overline{CN}$ are concurrent at point K and L, M, and N are points on $\overline{BC}$, $\overline{AC}$, and $\overline{AB}$, respectively. Points P, R, and Q are respective midpoints of $\overline{AL}$, $\overline{CN}$, and $\overline{BM}$. Prove that $\overline{DP}$, $\overline{EQ}$, and $\overline{FR}$ are concurrent if D, E, and F are respective midpoints of $\overline{BC}$, $\overline{AC}$, and $\overline{AB}$.

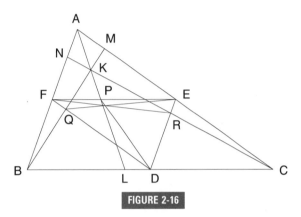

FIGURE 2-16

7. In △ABC (Figure 2-17), $\overline{AL}$, $\overline{BM}$, and $\overline{CN}$ are concurrent at point P. Points S, Q, and R are midpoints of $\overline{MN}$, $\overline{ML}$, and $\overline{NL}$, respectively. Prove that $\overrightarrow{AS}$, $\overrightarrow{BR}$, and $\overrightarrow{CQ}$ are concurrent.

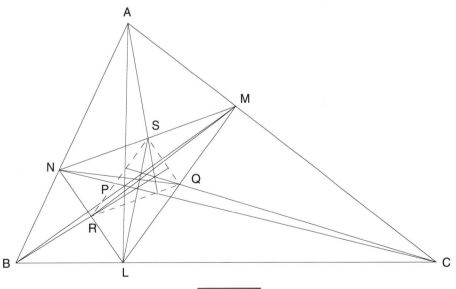

FIGURE 2-17

8. In $\triangle ABC$ (Figure 2-17), $\overline{AL}$, $\overline{BM}$, and $\overline{CN}$ are concurrent at point P. Points S, Q, and R are points on $\overline{MN}$, $\overline{ML}$, and $\overline{NL}$, respectively. If $\overline{LS}$, $\overline{MR}$, and $\overline{NQ}$ are concurrent, prove that $\overrightarrow{AS}$, $\overrightarrow{BR}$, and $\overrightarrow{CQ}$ are also concurrent.

9. Circles P, O, and Q are escribed circles of $\triangle ABC$, with the points of tangency indicated in Figure 2-18. Prove that $\overline{AD}$, $\overline{BE}$, and $\overline{CF}$ are concurrent.

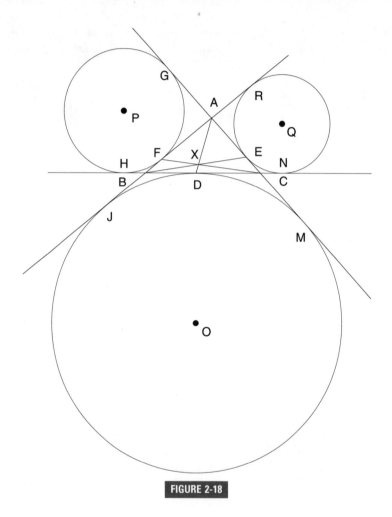

FIGURE 2-18

10. Given three circles, nonintersecting, mutually external, and with distinct radii, connect the intersection of internal common tangents of each pair of circles with the center of the other circle, as in Figure 2-19. Prove that the three resulting line segments are concurrent.

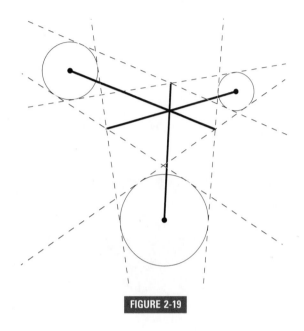

FIGURE 2-19

COLLINEARITY
OF POINTS

DUALITY

Many statements in geometry involve relationships between points and lines. In a statement concerning points and lines in a plane, when the word *point* is replaced by the word *line* and the word *line* is replaced by the word *point* each time these words are used in the statement, the newly formed statement is said to be the *dual* of the original statement. Occasionally, other modifications may need to be made in order to preserve proper sentence structure. This principle of duality was discovered by Charles Julien Brianchon (1785–1864) while using this relationship on a theorem by Blaise Pascal. We will visit these theorems later in this chapter. However, the transition from Chapter 2 to Chapter 3 follows the principle of duality because concurrency of lines is the dual of collinearity of points. The primary focus of this chapter is collinearity.

Let us first familiarize ourselves with the principle of duality. Consider the following examples of dual statements:

Statement	Dual Statement
1. Two distinct points determine a unique line.	1. Two distinct lines determine a unique point.
2. Any point contains an infinite number of lines.	2. Any line contains an infinite number of points.
3. Only one triangle is determined by three noncollinear points.	3. Only one trilateral is determined by three nonconcurrent lines.

This last example of duality demonstrates that related words also need to be changed when forming the dual of a statement. Specifically, note that *collinear* and *concurrent* are dual words, as are *triangle* and *trilateral*.

Recall Ceva's theorem (see Figure 3-1):

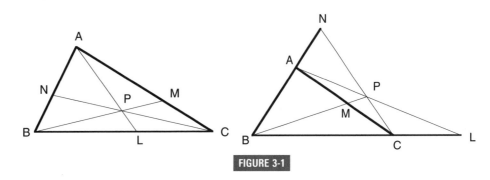

FIGURE 3-1

The three lines containing the vertices *A*, *B*, and *C* of △*ABC* and intersecting the opposite sides at points *L*, *M*, and *N*, respectively, are concurrent if and only if $\dfrac{AN}{NB} \cdot \dfrac{BL}{LC} \cdot \dfrac{CM}{MA} = 1$.

For the most part, the dual of a postulate is also a postulate, and the dual of a definition is itself a definition. Thus, if a statement is a theorem, its dual is likely to be a theorem as well.* In any case, we would at least have a statement that would be a good candidate to be a theorem. A valid proof would be needed to establish the statement as a theorem.

This is precisely what we will now investigate. With our knowledge of duality, we will form the dual statement of Ceva's theorem. Actually, it was the rediscovery of Menelaus of Alexandria's famous but forgotten theorem,[†] which we will discuss in the next section, that led Giovanni Ceva in the first book of his *De lineis rectis se invicem secantibus statica constructio* (Milan, 1678) to produce his theorem by the principle of duality. Note the duality relationship between the two theorems.

The three points *P*, *Q*, and *R* on the sides $\overleftrightarrow{AC}$, $\overleftrightarrow{AB}$, and $\overleftrightarrow{BC}$, respectively, of △*ABC* (see Figure 3-2) are collinear if and only if $\dfrac{AQ}{QB} \cdot \dfrac{BR}{RC} \cdot \dfrac{CP}{PA} = -1.$[‡]

This statement is in fact a theorem, known as *Menelaus's theorem,* that is the subject of the next section.

INTERACTIVE 3-2

Drag vertices *A*, *B*, and *C* to change the shape of the triangle; drag points *P* and *Q* and see that Menelaus's theorem is always true.

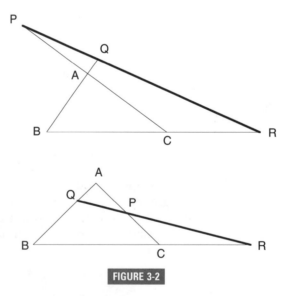

FIGURE 3-2

---

* In a geometric system exclusively based on postulates and definitions whose respective duals are all true, the dual of every theorem is also a theorem. This claim is easily justified by realizing that the proof of the dual of a theorem can be produced by simply replacing each statement of the proof of the original theorem by its dual.

† During the Dark Ages, much of classical Greek mathematics was lost and forgotten.

‡ The reason for the negative sign is explained in the proof of this theorem.

MENELAUS'S THEOREM

Menelaus of Alexandria, in about A.D. 100 in a work titled *Sphaerica*, produced the well-known plane version of the theorem that we will present here. He used the plane version to develop the spherical analogue,* which was the purpose of his treatise. As we have mentioned, this theorem, which today bears Menelaus's name, did not become popular until it was rediscovered by Giovanni Ceva as a part of his work in 1678.

THEOREM 3.1 (**Menelaus's theorem**) The three points P, Q, and R on the sides $\overleftrightarrow{AC}$, $\overleftrightarrow{AB}$, and $\overleftrightarrow{BC}$, respectively, of $\triangle ABC$ are collinear if and only if $\dfrac{AQ}{QB} \cdot \dfrac{BR}{RC} \cdot \dfrac{CP}{PA} = -1.$

Like Ceva's theorem, Menelaus's theorem is an equivalence and therefore requires proofs for each of the two statements (converses of each other) that comprise the entire theorem. We will first prove that *if the three points P, Q, and R on the sides $\overleftrightarrow{AC}$, $\overleftrightarrow{AB}$, and $\overleftrightarrow{BC}$,, respectively, of $\triangle ABC$ are collinear, then* $\dfrac{AQ}{QB} \cdot \dfrac{BR}{RC} \cdot \dfrac{CP}{PA} = -1.$ We offer two proofs of this part of Menelaus's theorem.

❶roof I Draw a line containing C, parallel to $\overleftrightarrow{AB}$ and intersecting $\overline{PQR}$ (or $\overline{QPR}$) at point D (see Figure 3-3).

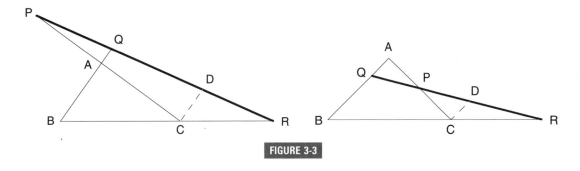

FIGURE 3-3

$$\triangle DCR \sim \triangle QBR \Rightarrow \frac{DC}{QB} = \frac{RC}{BR} \quad or \quad DC = \frac{(QB)(RC)}{BR} \qquad (I)$$

$$\triangle PDC \sim \triangle PQA \Rightarrow \frac{DC}{AQ} = \frac{CP}{PA} \quad or \quad DC = \frac{(AQ)(CP)}{PA} \qquad (II)$$

---

* The spherical analogue to Theorem 3.1 for spherical triangle ABC is
$\dfrac{\sin \widehat{AQ}}{\sin \widehat{QB}} \cdot \dfrac{\sin \widehat{BR}}{\sin \widehat{RC}} \cdot \dfrac{\sin \widehat{CP}}{\sin \widehat{PA}} = -1.$

From (I) and (II), we get:

$$\frac{(QB)(RC)}{BR} = \frac{(AQ)(CP)}{PA} \quad \text{or} \quad (QB)(RC)(PA) = (AQ)(CP)(BR)$$

This gives us:

$$\frac{AQ}{QB} \cdot \frac{BR}{RC} \cdot \frac{CP}{PA} = 1$$

By taking direction into account in the left-hand diagram of Figure 3-3, we see that $\frac{AQ}{QB}, \frac{BR}{RC}$, and $\frac{CP}{PA}$ are each negative ratios; in the right-hand diagram of Figure 3-3, $\frac{BR}{RC}$ is a negative ratio, whereas $\frac{AQ}{QB}$ and $\frac{CP}{PA}$ are positive ratios. Because in each case there is an odd number of negative ratios:

$$\frac{AQ}{QB} \cdot \frac{BR}{RC} \cdot \frac{CP}{PA} = -1 \; \bullet$$

℗roof II Once again we begin by assuming collinearity of P, Q, and R. Draw $\overline{BM} \perp \overleftrightarrow{PR}$, $\overline{AN} \perp \overleftrightarrow{PR}$, and $\overline{CL} \perp \overleftrightarrow{PR}$ (see Figure 3-4).

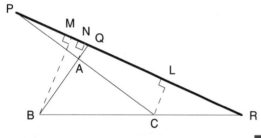 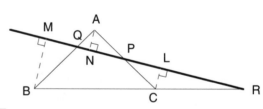

FIGURE 3-4

$$\triangle BMQ \sim \triangle ANQ \Rightarrow \frac{AQ}{QB} = \frac{AN}{BM} \tag{I}$$

$$\triangle LCP \sim \triangle NAP \Rightarrow \frac{CP}{PA} = \frac{LC}{AN} \tag{II}$$

$$\triangle MRB \sim \triangle LRC \Rightarrow \frac{BR}{RC} = \frac{BM}{LC} \tag{III}$$

By multiplying (I), (II), and (III), we get, numerically:

$$\frac{AQ}{QB} \cdot \frac{CP}{PA} \cdot \frac{BR}{RC} = \frac{AN}{BM} \cdot \frac{LC}{AN} \cdot \frac{BM}{LC} = 1$$

In the left-hand diagram of Figure 3-4, $\dfrac{AQ}{QB}$ is negative, $\dfrac{CP}{PA}$ is negative, and $\dfrac{BR}{RC}$ is negative. Therefore:

$$\frac{AQ}{QB} \cdot \frac{CP}{PA} \cdot \frac{BR}{RC} = -1$$

In the right-hand diagram of Figure 3-4, $\dfrac{AQ}{QB}$ is positive, $\dfrac{CP}{PA}$ is positive, and $\dfrac{BR}{RC}$ is negative. Therefore:

$$\frac{AQ}{QB} \cdot \frac{CP}{PA} \cdot \frac{BR}{RC} = -1 \; \bullet$$

To complete the proof of Menelaus's theorem, we must now prove the *converse* of the theorem we just proved. We will prove that *if the three points P, Q, and R are on the sides $\overleftrightarrow{AC}$, $\overleftrightarrow{AB}$, and $\overleftrightarrow{BC}$, respectively, and if* $\dfrac{AQ}{QB} \cdot \dfrac{BR}{RC} \cdot \dfrac{CP}{PA} = -1$, *then points P, Q, and R are collinear.*

℗roof   In Figure 3-2, let the line containing points R and Q intersect $\overline{AC}$ at P'. Using the portion of the theorem just proved, we know that:

$$\frac{AQ}{QB} \cdot \frac{BR}{RC} \cdot \frac{CP'}{P'A} = -1$$

However, our hypothesis tells us that:

$$\frac{AQ}{QB} \cdot \frac{BR}{RC} \cdot \frac{CP}{PA} = -1$$

Therefore $\dfrac{CP'}{P'A} = \dfrac{CP}{PA}$, which indicates that P and P' must coincide. This proves the collinearity. $\bullet$

Menelaus's theorem provides us with a useful method for proving points collinear.

APPLICATIONS OF MENELAUS'S THEOREM

Before investigating other famous theorems that can be proved using Menelaus's theorem, we will consider a few applications of Menelaus's theorem. Each of these unnamed theorems presents some very interesting results that are very easily proved using Menelaus's theorem.

Application 1 Prove that the interior angle bisectors of two angles of a nonisosceles triangle and the exterior angle bisector of the third angle meet the opposite sides in three collinear points. ●

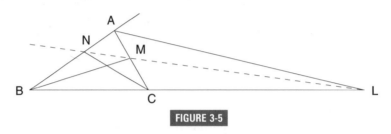

FIGURE 3-5

Proof In $\triangle ABC$, $\overline{BM}$ and $\overline{CN}$ are the interior angle bisectors, while $\overline{AL}$ bisects the exterior angle at point A (see Figure 3-5). Because the bisector of an angle (interior or exterior) of a triangle partitions the opposite side proportionally to the two remaining sides, we have:

$$\frac{AM}{MC} = \frac{AB}{BC} \quad \frac{BN}{NA} = \frac{BC}{AC} \quad \text{and} \quad \frac{CL}{BL} = \frac{AC}{AB}$$

Therefore, by multiplication:

$$\frac{AM}{MC} \cdot \frac{BN}{NA} \cdot \frac{CL}{BL} = \frac{AB}{BC} \cdot \frac{BC}{AC} \cdot \frac{AC}{AB} = 1$$

However, because $\dfrac{CL}{BL} = \dfrac{-CL}{LB}$:

$$\frac{AM}{MC} \cdot \frac{BN}{NA} \cdot \frac{CL}{LB} = -1$$

Thus, by Menelaus's theorem, points N, M, and L must be collinear. ●

Application 2 Prove that the exterior angle bisectors of any nonisosceles triangle meet the opposite sides in three collinear points. ●

Proof   In $\triangle ABC$, the bisectors of the exterior angles at A, B, and C meet the opposite sides (extended) at points N, L, and M, respectively (see Figure 3-6). Because the bisector of an angle (interior or exterior) of a triangle partitions the opposite side proportionally to the two remaining sides, we have:

$$\frac{CL}{AL} = \frac{BC}{AB} \quad \frac{AM}{BM} = \frac{AC}{BC} \quad \text{and} \quad \frac{BN}{CN} = \frac{AB}{AC}$$

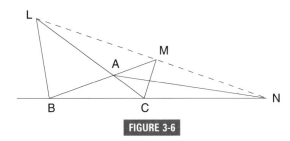

FIGURE 3-6

Therefore:

$$\frac{CL}{AL} \cdot \frac{AM}{BM} \cdot \frac{BN}{CN} = \frac{BC}{AB} \cdot \frac{AC}{BC} \cdot \frac{AB}{AC} = -1 \text{ (because all three ratios are negative)}$$

Thus, by Menelaus's theorem, points L, M, and N are collinear. ●

Application 3   A circle through vertices B and C of $\triangle ABC$ meets $\overline{AB}$ at point P and $\overline{AC}$ at point R. If $\overleftrightarrow{PR}$ meets $\overleftrightarrow{BC}$ at point Q, prove that $\frac{QC}{QB} = \frac{(RC)(AC)}{(PB)(AB)}$. ●

Proof   Consider $\triangle ABC$ with transversal $\overline{QPR}$ (Figure 3-7). By Menelaus's theorem:

$$\frac{RC}{AR} \cdot \frac{AP}{PB} \cdot \frac{QB}{CQ} = -1$$

Then, considering absolute values, we have:

$$\frac{QC}{QB} = \frac{RC}{AR} \cdot \frac{AP}{PB} \tag{I}$$

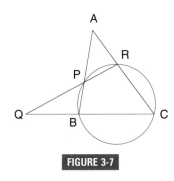

FIGURE 3-7

However, $(AP)(AB) = (AR)(AC)$. (If two secant segments intersect outside the circle, the product of the lengths of one secant segment and its external segment equals the product of the lengths of the other secant segment and its external segment.) Therefore:

$$\frac{AP}{AR} = \frac{AC}{AB} \tag{II}$$

By substituting (II) into (I), we get our desired result:

$$\frac{QC}{QB} = \frac{(RC)(AC)}{(PB)(AB)} \quad ●$$

Often, both Menelaus's theorem and its dual, Ceva's theorem, are needed to solve a problem or prove a theorem. The next applications demonstrate this.

Application 4

In right triangle ABC, points P and Q are on $\overline{BC}$ and $\overline{AC}$, respectively, such that $CP = CQ = 2$. Through R, the point of intersection of $\overline{AP}$ and $\overline{BQ}$, a line is drawn also passing through point C and meeting $\overline{AB}$ at point S. $\overleftrightarrow{PQ}$ meets $\overleftrightarrow{AB}$ at point T. If hypotenuse $AB = 10$ and $AC = 8$, find TS (Figure 3-8). ●

Solution

In right triangle ABC, hypotenuse $AB = 10$ and $AC = 8$, so $BC = 6$ (by the Pythagorean theorem). In $\triangle ABC$, because $\overline{AP}$, $\overline{BQ}$, and $\overline{CS}$ are concurrent, by Ceva's theorem:

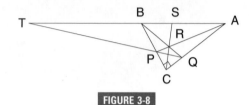

FIGURE 3-8

$$\frac{AQ}{QC} \cdot \frac{CP}{PB} \cdot \frac{BS}{SA} = 1$$

Substituting, we get:

$$\frac{6}{2} \cdot \frac{2}{4} \cdot \frac{BS}{10 - BS} = 1 \quad \text{or} \quad BS = 4$$

Now consider $\triangle ABC$ with transversal $\overline{QPT}$. By Menelaus's theorem:

$$\frac{AQ}{QC} \cdot \frac{CP}{PB} \cdot \frac{BT}{TA} = -1$$

Because we are not dealing with directed line segments, this may be restated as:

$$(AQ)(CP)(BT) = (QC)(PB)(AT)$$

Substituting, we get:

$$(6)(2)(BT) = (2)(4)(BT + 10)$$

Solving for BT gives us $BT = 20$, and thus $TS = 24$. ●

Application 5

In quadrilateral $ABCD$, $\overleftrightarrow{AB}$ and $\overleftrightarrow{CD}$ meet at point P, while $\overleftrightarrow{AD}$ and $\overleftrightarrow{BC}$ meet at point Q. Diagonals $\overleftrightarrow{AC}$ and $\overleftrightarrow{BD}$ meet $\overleftrightarrow{PQ}$ at points X and Y, respectively. Prove that $\dfrac{PX}{XQ} = -\dfrac{PY}{YQ}$ (see Figure 3-9). ●

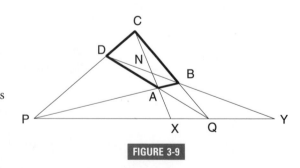

FIGURE 3-9

℗roof   Consider $\triangle PQC$ with $\overline{PB}$, $\overline{QD}$, and $\overline{CX}$ concurrent. By Ceva's theorem:

$$\frac{PX}{XQ}\cdot\frac{QB}{BC}\cdot\frac{CD}{DP}=1 \qquad\text{(I)}$$

Now consider $\triangle PQC$ with $\overline{DBY}$ as a transversal. By Menelaus's theorem:

$$\frac{PY}{YQ}\cdot\frac{QB}{BC}\cdot\frac{CD}{DP}=-1 \qquad\text{(II)}$$

Therefore, from (I) and (II):

$$\frac{PX}{XQ}=-\frac{PY}{YQ}\;\bullet$$

We will now consider some rather famous theorems that can be proved using Menelaus's theorem.

DESARGUES'S THEOREM

During his lifetime, Gérard Desargues (1591–1661) did not enjoy the important stature as a mathematician that he has attained in later years. This lack of popularity was in part due to the then recent development of analytic geometry by René Descartes (1596–1650) and to Desargues's introduction of many new and largely unfamiliar terms. (Incidentally, we make every effort in this book not to introduce any new terms in order to make it more reader-friendly—we want to learn from Desargues's misjudgment.)

In 1648, his pupil, Abraham Bosse, a master engraver, published a book titled *Manière universelle de M. Desargues, pour pratiquer la perspective,* which was not popularized until about two centuries later. This book contained a theorem that in the nineteenth century became one of the fundamental propositions of projective geometry. It is this theorem that is of interest to us here. It involves placing any two triangles in a position that will enable the three lines joining corresponding vertices to be concurrent. Remarkably, when this is achieved the pairs of corresponding sides meet in three collinear points. We will prove Desargues's theorem by using Menelaus's theorem.

THEOREM 3.2   **(Desargues's theorem)**   If $\triangle A_1B_1C_1$ and $\triangle A_2B_2C_2$ are situated so that the lines joining the corresponding vertices, $\overleftrightarrow{A_1A_2}$, $\overleftrightarrow{B_1B_2}$, and $\overleftrightarrow{C_1C_2}$, are concurrent, then the pairs of corresponding sides intersect in three collinear points.

INTERACTIVE 3-10

Drag vertices A_1, B_1, C_1 and A_2, B_2, C_2 to change the shape of the triangles and point P and see that Desargues's theorem is true.

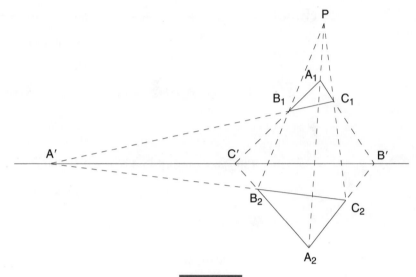

FIGURE 3-10

Proof   In Figure 3-10, $\overleftrightarrow{A_1A_2}$, $\overleftrightarrow{B_1B_2}$, and $\overleftrightarrow{C_1C_2}$ all meet at point P, by hypothesis. $\overleftrightarrow{B_2C_2}$ and $\overleftrightarrow{B_1C_1}$ meet at A′; $\overleftrightarrow{A_2C_2}$ and $\overleftrightarrow{A_1C_1}$ meet at B′; $\overleftrightarrow{B_2A_2}$ and $\overleftrightarrow{B_1A_1}$ meet at C′.

Consider $\overline{A'B_1C_1}$ to be a transversal of $\triangle PB_2C_2$. Therefore, by Menelaus's theorem:

$$\frac{PB_1}{B_1B_2} \cdot \frac{B_2A'}{A'C_2} \cdot \frac{C_2C_1}{C_1P} = -1 \qquad\qquad \text{(I)}$$

Similarly, considering $\overline{C'B_1A_1}$ as a transversal of $\triangle PB_2A_2$:

$$\frac{PA_1}{A_1A_2} \cdot \frac{A_2C'}{C'B_2} \cdot \frac{B_2B_1}{B_1P} = -1 \quad \text{(Menelaus's theorem)} \qquad \text{(II)}$$

Now, taking $\overline{B'A_1C_1}$ as a transversal of $\triangle PA_2C_2$:

$$\frac{PC_1}{C_1C_2} \cdot \frac{C_2B'}{B'A_2} \cdot \frac{A_2A_1}{A_1P} = -1 \quad \text{(Menelaus's theorem)} \qquad \text{(III)}$$

By multiplying (I), (II), and (III), we get:

$$\frac{B_2A'}{A'C_2} \cdot \frac{A_2C'}{C'B_2} \cdot \frac{C_2B'}{B'A_2} = -1$$

Thus, by Menelaus's theorem, applied to $\triangle A_2B_2C_2$, we have points A', B', and C' collinear. ●

It should be noted that the converse of Desargues's theorem is also true. It is the dual of the original theorem. We leave the proof as an exercise.

To appreciate the value of Desargues's theorem, we will examine some applications. Although each may be proved in other ways, we will use the method employing Desargues's theorem.

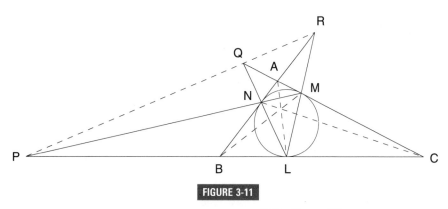

FIGURE 3-11

Ⓐpplication 6   A circle inscribed in △*ABC* is tangent to sides $\overline{BC}$, $\overline{CA}$, and $\overline{AB}$ at points *L*, *M*, and *N*, respectively. $\overleftrightarrow{MN}$ intersects $\overleftrightarrow{BC}$ at point *P*, $\overleftrightarrow{NL}$ intersects $\overleftrightarrow{AC}$ at point *Q*, and $\overleftrightarrow{ML}$ intersects $\overleftrightarrow{AB}$ at point *R*. Prove that points *P*, *Q*, and *R* are collinear (see Figure 3-11). ●

Ⓟroof   Because the tangent segments from an external point to a circle are congruent:

$$AN = AM \quad NB = BL \quad MC = LC$$

Therefore:

$$\frac{AN}{NB} \cdot \frac{BL}{LC} \cdot \frac{MC}{AM} = 1$$

By Ceva's theorem, $\overleftrightarrow{AL}$, $\overleftrightarrow{BM}$, and $\overleftrightarrow{CN}$ are concurrent. Because these are the lines joining the corresponding vertices of △*ABC* and △*LMN*, by Desargues's theorem the intersections of the corresponding sides are collinear; therefore points *P*, *Q*, and *R* are collinear. ●

Ⓐpplication 7   In △*ABC*, points *F*, *E*, and *D* are the feet of the altitudes drawn from the vertices *A*, *B*, and *C*, respectively. The sides of pedal* triangle *FED*, $\overleftrightarrow{EF}$, $\overleftrightarrow{DF}$, and $\overleftrightarrow{DE}$, intersect the sides of △*ABC*, $\overleftrightarrow{AB}$, $\overleftrightarrow{AC}$, and $\overleftrightarrow{BC}$, at points *M*, *N*, and *L*, respectively. Prove that points *M*, *N*, and *L* are collinear (see Figure 3-12). ●

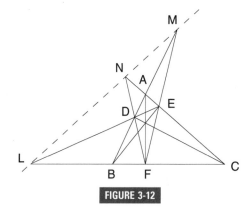

FIGURE 3-12

Ⓟroof   Let *A*, *B*, *C* and *F*, *E*, *D* be corresponding vertices of △*ABC* and △*FED*. Because $\overline{AF}$, $\overline{CD}$, and $\overline{BE}$ are concurrent (they are the altitudes of △*ABC*), the intersections of the corresponding sides $\overleftrightarrow{DE}$ and $\overleftrightarrow{BC}$, $\overleftrightarrow{FE}$ and $\overleftrightarrow{BA}$, and $\overleftrightarrow{FD}$ and $\overleftrightarrow{CA}$ are collinear by Desargues's theorem. ●

---

* A pedal triangle (of a given triangle) is formed by joining the feet of the perpendiculars from any given point to the sides of the given triangle.

PASCAL'S THEOREM

Blaise Pascal (1623–1662), a contemporary of Desargues, is regarded today as one of the true geniuses in the history of mathematics. Although eccentricities kept him from achieving his true potential, he is considered one of the originators of the formalized study of probability (an outgrowth of his correspondences with Fermat), and he made many important contributions to other branches of mathematics. We concern ourselves here with one of his contributions to geometry.

In 1640, at the age of sixteen, Pascal published a one-page paper titled *Essay pour les coniques*. It contained a theorem that Pascal referred to as *mysterium hexagrammicum*. The work highly impressed Descartes, who couldn't believe it was the work of a boy. This theorem states that the intersections of the opposite sides of a hexagon inscribed in a conic section are collinear. For our purposes, we will consider only the case in which the conic section is a circle and the hexagon has no pair of opposite sides parallel.

THEOREM 3.3   **(Pascal's theorem)**   If a hexagon with no pair of opposite sides parallel is inscribed in a circle, then the intersections of the opposite sides are collinear.

Proof   Hexagon $ABCDEF$ is inscribed in a circle (see Figure 3-13). The pairs of opposite sides $\overleftrightarrow{AB}$ and $\overleftrightarrow{DE}$ meet at point L, $\overleftrightarrow{CB}$ and $\overleftrightarrow{EF}$ meet at point M, and $\overleftrightarrow{CD}$ and $\overleftrightarrow{AF}$ meet at point N. Also, $\overleftrightarrow{AB}$ meets $\overleftrightarrow{CN}$ at point X, $\overleftrightarrow{EF}$ meets $\overleftrightarrow{CN}$ at point Y, and $\overleftrightarrow{EF}$ meets $\overleftrightarrow{AB}$ at point Z.

Consider $\overleftrightarrow{BC}$ to be a transversal of $\triangle XYZ$. Then, by Menelaus's theorem:

$$\frac{ZB}{BX} \cdot \frac{XC}{CY} \cdot \frac{YM}{MZ} = -1 \qquad \text{(I)}$$

Taking $\overleftrightarrow{AF}$ to be a transversal of $\triangle XYZ$:

$$\frac{ZA}{AX} \cdot \frac{YF}{FZ} \cdot \frac{XN}{NY} = -1 \quad \text{(Menelaus's theorem)} \qquad \text{(II)}$$

Also, because $\overleftrightarrow{DE}$ is a transversal of $\triangle XYZ$:

$$\frac{XD}{DY} \cdot \frac{YE}{EZ} \cdot \frac{ZL}{LX} = -1 \quad \text{(Menelaus's theorem)} \qquad \text{(III)}$$

By multiplying (I), (II), and (III), we get:

$$\frac{YM}{MZ} \cdot \frac{XN}{NY} \cdot \frac{ZL}{LX} \cdot \frac{(ZB)(ZA)}{(EZ)(FZ)} \cdot \frac{(XD)(XC)}{(AX)(BX)} \cdot \frac{(YE)(YF)}{(DY)(CY)} = -1 \qquad \text{(IV)}$$

When two secant segments are drawn to a circle from an external point, the product of the lengths of one secant and its external segment equals the product of the lengths of the other secant and its external segment.

INTERACTIVE 3-13

Drag points *A, B, C, D, E,* and *F* on the circle to see that Pascal's theorem is true.

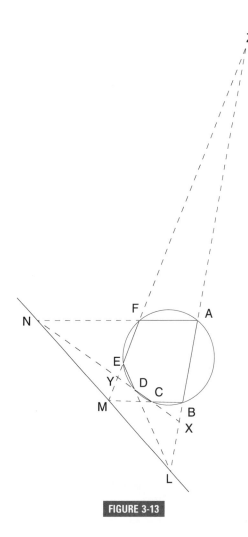

FIGURE 3-13

Therefore:

$$\frac{(ZB)(ZA)}{(EZ)(FZ)} = 1 \qquad\qquad\qquad \text{(V)}$$

$$\frac{(XD)(XC)}{(AX)(BX)} = 1 \qquad\qquad\qquad \text{(VI)}$$

$$\frac{(YE)(YF)}{(DY)(CY)} = 1 \qquad\qquad\qquad \text{(VII)}$$

By substituting (V), (VI), and (VII) into (IV), we get:

$$\frac{YM}{MZ} \cdot \frac{XN}{NY} \cdot \frac{ZL}{LX} = -1$$

Thus, by Menelaus's theorem, points *M, N,* and *L* must be collinear. ●

It is interesting to note that Pascal's theorem can be extended in the following manner.

| **THEOREM 3.4** | **(variation on Pascal's theorem)**   If a hexagon has its vertices on a circle *in any order,* then the intersections (if they exist) of the opposite sides are collinear. |

As an example of this variation, you are invited to follow the proof of Theorem 3.3 using the diagram in Figure 3-14. Only one minor adjustment needs to be made, and that is the reason for equations (V) through (VII). Note that the same pairs of "opposite sides" are used here as were used earlier.

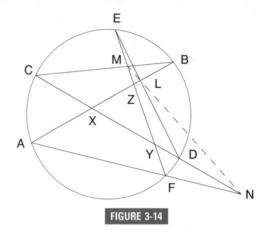

FIGURE 3-14

Pascal's theorem has many applications. We will consider only a few.

Application 8   Point P is any point in the interior of $\triangle ABC$. Points M and N are the feet of the perpendiculars from point P to $\overline{AB}$ and $\overline{AC}$, respectively. $\overline{AK} \perp \overleftrightarrow{CP}$ at point K, and $\overline{AL} \perp \overleftrightarrow{BP}$ at point L (see Figure 3-15). Prove that $\overleftrightarrow{KM}$, $\overleftrightarrow{LM}$, and $\overleftrightarrow{BC}$ are concurrent. ●

Proof   We can easily prove that some points A, K, M, P, N, and L all lie on the circle with diameter $\overline{AP}$. We can justify this by realizing that right angles AKP and AMP are inscribed in the same semicircle, as is the case for right angles ALP and ANP. Using the variation on Pascal's theorem (Theorem 3.4), we notice that for inscribed hexagon $AKMPNL$, the pairs of sides intersect as follows:

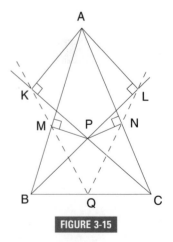

FIGURE 3-15

$$\overleftrightarrow{AM} \cap \overleftrightarrow{LP} = B$$
$$\overleftrightarrow{AN} \cap \overleftrightarrow{KP} = C$$
$$\overleftrightarrow{KM} \cap \overleftrightarrow{LN} = Q$$

By Pascal's theorem, points B, C, and Q are collinear, which is to say that $\overleftrightarrow{KM}$, $\overleftrightarrow{LM}$, and $\overleftrightarrow{BC}$ are concurrent. ●

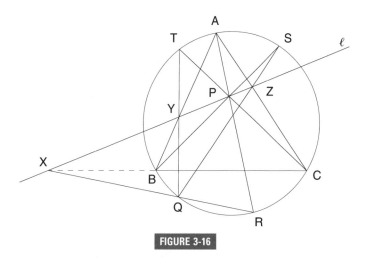

FIGURE 3-16

Application 9 Select any point P not on $\triangle ABC$ and a line ℓ containing P and intersecting sides $\overleftrightarrow{BC}$, $\overleftrightarrow{AB}$, and $\overleftrightarrow{AC}$ at points X, Y, and Z, respectively. Let $\overleftrightarrow{AP}$, $\overleftrightarrow{BP}$, and $\overleftrightarrow{CP}$ intersect the circumcircle of $\triangle ABC$ at points R, S, and T, respectively (see Figure 3-16). Prove that $\overleftrightarrow{RX}$, $\overleftrightarrow{SZ}$, and $\overleftrightarrow{TY}$ are concurrent. ●

Proof Let $\overleftrightarrow{RX}$ intersect the circumcircle at point Q. Consider hexagon $ARQTCB$ and apply Pascal's theorem to it. We notice that because $\overleftrightarrow{AR} \cap \overleftrightarrow{AB}$ at point P and $\overleftrightarrow{RQ} \cap \overleftrightarrow{CB}$ at point X, $\overleftrightarrow{TQ} \cap \overleftrightarrow{AB}$ at a point on ℓ, which must be Y (because $\overleftrightarrow{AB} \cap \ell$ at point Y).

Now consider hexagon $ARQSBC$. Because $\overleftrightarrow{AR} \cap \overleftrightarrow{SB}$ at P and $\overleftrightarrow{RQ} \cap \overleftrightarrow{CB}$ at X, $\overleftrightarrow{SQ} \cap \overleftrightarrow{AC}$ at a point on ℓ, which must be Z. Thus $\overleftrightarrow{RX}$, $\overleftrightarrow{SZ}$, and $\overleftrightarrow{TY}$ are and concurrent. ●

BRIANCHON'S THEOREM

In 1806, at the age of twenty-one, a student at the École Polytechnique, Charles Julien Brianchon (1785–1864), published an article in the *Journal de L'École Polytechnique* that was to become one of the fundamental contributions to the study of conic sections in projective geometry. His development led to a restatement of the somewhat forgotten theorem of Pascal and its extension, after which Brianchon stated a new theorem that now bears his name. Brianchon's theorem, which states "In any hexagon circumscribed about a conic section, the three diagonals cross each other in the same point,"* bears a curious

---

* *Source Book in Mathematics,* edited by D. E. Smith (New York: McGraw-Hill Book Co., 1929), p. 336.

resemblance to Pascal's theorem. They are, in fact, duals of each other. This can be easily seen by comparing the following versions of each theorem:

Pascal's Theorem
The <u>points of intersection</u> of the opposite <u>sides</u> of a hexagon <u>inscribed in</u> a conic section are <u>collinear</u>.

Brianchon's Theorem
The <u>lines joining</u> the opposite <u>vertices</u> of a hexagon <u>circumscribed about</u> a conic section are <u>concurrent</u>.

Notice that the two statements above are alike except for the underlined words, which are duals of one another. As with Pascal's theorem, we will consider only the conic section that is a circle.

THEOREM 3.5 **(Brianchon's theorem)** If a hexagon is circumscribed about a circle (see Figure 3-17), the lines containing opposite vertices are concurrent.

Drag points *A, B, C, D, E,* and *F* on the circle and see that Brianchon's theorem is true.

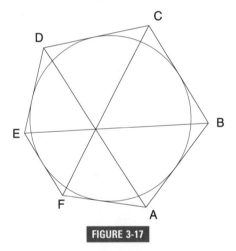

FIGURE 3-17

The simplest proofs of this theorem require knowledge of concepts from projective geometry. Although we are prepared at this point to prove this theorem by Euclidean methods, our proof will be more concise if we wait until we study radical axes later in this chapter on page 71.

Brianchon suggested the following application immediately after the statement of his new theorem.

Application 10   Pentagon *ABCDE* is circumscribed about a circle, with points of tangency at *F*, *M*, *N*, *R*, and *S*. If diagonals $\overline{AD}$ and $\overline{BE}$ intersect at point *P*, prove that points *C*, *P*, and *F* are collinear (see Figure 3-18). ●

Proof   Consider the hexagon circumscribed about a circle (Figure 3-17) having its sides $\overline{AF}$ and $\overline{EF}$ merge into one line segment. Thus $\overline{AFE}$ is now a side of a circumscribed pentagon with *F* as one point of tangency (see Figure 3-18). Thus we can view the pentagon in Figure 3-18 as a degenerate hexagon. We then simply apply Brianchon's theorem to this degenerate hexagon to obtain our desired conclusion. That is, $\overline{AD}$, $\overline{BE}$, and $\overline{CF}$ are concurrent at point *P*, or points *C*, *P*, and *F* are collinear. ●

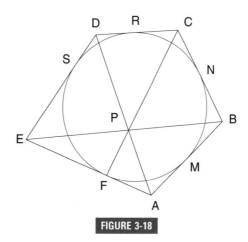

FIGURE 3-18

PAPPUS'S THEOREM

Consider the vertices of a hexagon $AB'CA'BC'$ (Figure 3-19) located alternately on two lines (see Figure 3-20). Suppose we now draw the lines that were the opposite sides of the hexagon to locate their point of intersection. We find that the three points of intersection of these pairs of "opposite sides" are collinear. This conclusion was first published by Pappus of Alexandria in his *Mathematical Collection* circa A.D. 300.

For the purpose of providing a proof, we will restate Pappus's theorem. You will notice once again that the proof uses Menelaus's theorem repeatedly.

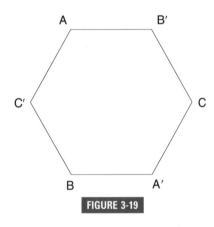

FIGURE 3-19

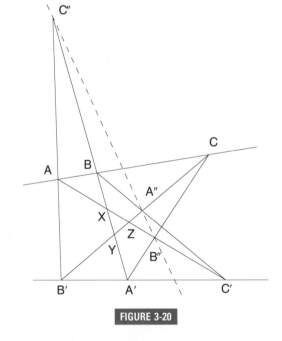

FIGURE 3-20

INTERACTIVE 3-20

Drag points A', B', C', A, B, and C and see that Pappus's theorem is true.

THEOREM 3.6 **(Pappus's theorem)** Points A, B, and C are on one line, and points A', B', and C' are on another line (in any order). If $\overleftrightarrow{AB'}$ and $\overleftrightarrow{A'B}$ meet at C'', $\overleftrightarrow{AC'}$ and $\overleftrightarrow{A'C}$ meet at B'', and $\overleftrightarrow{BC'}$ and $\overleftrightarrow{B'C}$ meet at A'', then points A'', B'', and C'' are collinear.

Proof In Figure 3-20, $\overline{B'C}$ meets $\overline{A'B}$ at point Y, $\overline{AC'}$ meets $\overline{A'B}$ at point X, and $\overline{B'C}$ meets $\overline{AC'}$ at point Z. Consider $\overline{C''AB'}$ as a transversal of $\triangle XYZ$. By Menelaus's theorem:

$$\frac{ZB'}{YB'} \cdot \frac{XA}{ZA} \cdot \frac{YC''}{XC''} = -1 \tag{I}$$

Taking $\overline{A'B''C}$ as a transversal of $\triangle XYZ$:

$$\frac{YA'}{XA'} \cdot \frac{XB''}{ZB''} \cdot \frac{ZC}{YC} = -1 \text{ (Menelaus's theorem)} \tag{II}$$

$\overline{BA''C'}$ is also a transversal of $\triangle XYZ$, so:

$$\frac{YB}{XB} \cdot \frac{ZA''}{YA''} \cdot \frac{XC'}{ZC'} = -1 \text{ (Menelaus's theorem)} \qquad \text{(III)}$$

Multiplying (I), (II), and (III) gives us:

$$\frac{YC''}{XC''} \cdot \frac{XB''}{ZB''} \cdot \frac{ZA''}{YA''} \cdot \frac{ZB'}{YB'} \cdot \frac{YA'}{XA'} \cdot \frac{XC'}{ZC'} \cdot \frac{XA}{ZA} \cdot \frac{ZC}{YC} \cdot \frac{YB}{XB} = -1 \qquad \text{(IV)}$$

Because points A, B, and C are collinear and points A', B', and C' are collinear, we obtain the following two relationships by Menelaus's theorem (when we consider each line as a transversal of $\triangle XYZ$):

$$\frac{ZB'}{YB'} \cdot \frac{YA'}{XA'} \cdot \frac{XC'}{ZC'} = -1 \qquad \text{(V)}$$

$$\frac{XA}{ZA} \cdot \frac{ZC}{YC} \cdot \frac{YB}{XB} = -1 \qquad \text{(VI)}$$

Substituting (V) and (VI) into (IV), we get:

$$\frac{YC''}{XC''} \cdot \frac{XB''}{ZB''} \cdot \frac{ZA''}{YA''} = -1$$

Thus, by Menelaus's theorem, points A'', B'', and C'' are collinear. ●

THE SIMSON LINE

One of the great injustices in the history of mathematics involves a theorem originally published by William Wallace (1768–1843) in Thomas Leybourn's *Mathematical Repository* (1799), which through careless misquotes has been attributed to Robert Simson (1687–1768), a famous English interpreter of Euclid's *Elements*. (See pages 96–97 for more on Simson.) We will use the popular reference *Simson's theorem* throughout this book.

▌THEOREM 3.7 (**Simson's theorem**) The feet of the perpendiculars drawn from any point on the circumcircle of a triangle to the sides of the triangle are collinear.

In Figure 3-21, point P is on the circumcircle of $\triangle ABC$. $\overleftrightarrow{PY} \perp \overleftrightarrow{AC}$ at point Y, $\overleftrightarrow{PZ} \perp \overleftrightarrow{AB}$ at point Z, and $\overleftrightarrow{PX} \perp \overleftrightarrow{BC}$ at point X. According to Simson's (i.e., Wallace's) theorem, points X, Y, and Z are collinear. This line is usually referred to as the *Simson line* (sometimes called the *pedal line*).

Although not necessarily the simplest proof of Simson's theorem, for the sake of consistency we will prove this theorem using Menelaus's theorem. We will provide a second method of proof to demonstrate this theorem's independence.

INTERACTIVE 3-21

Drag vertices *A, B,* and *C* to change the shape of the triangle; drag point *P* on the circle and watch the Simson line.

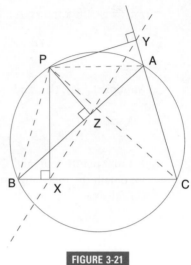

FIGURE 3-21

Proof I   (See Figure 3-21.) Draw $\overline{PA}$, $\overline{PB}$, and $\overline{PC}$.

$$m\angle PBA = \frac{1}{2} m \stackrel{\frown}{AP}$$

$$m\angle PCA = \frac{1}{2} m \stackrel{\frown}{AP}$$

Therefore $m\angle PBA = m\angle PCA = a$. Thus:

$$\frac{BZ}{PZ} = \cot a = \frac{CY}{PY} \text{ (in } \triangle PZB \text{ and } \triangle PYC) \quad \text{or} \quad \frac{BZ}{PZ} = \frac{CY}{PY}$$

This implies:

$$\frac{BZ}{CY} = \frac{PZ}{PY} \tag{I}$$

Similarly, $m\angle PAB = m\angle PCB = b$ (both are $\frac{1}{2}m\stackrel{\frown}{PB}$). Therefore:

$$\frac{AZ}{PZ} = \cot b = \frac{CX}{PX} \text{ (in } \triangle PAZ \text{ and } \triangle PCX) \quad \text{or} \quad \frac{AZ}{PZ} = \frac{CX}{PX}$$

This implies:

$$\frac{CX}{AZ} = \frac{PX}{PZ} \tag{II}$$

Because $\angle PBC$ and $\angle PAC$ are opposite angles of an inscribed (cyclic) quadrilateral, they are supplementary. However, $\angle PAY$ is also supplementary to $\angle PAC$. Therefore:

$$m\angle PBC = m\angle PAY = c$$

Thus:

$$\frac{BX}{PX} = \cot c = \frac{AY}{PY} \text{ (in } \triangle PBX \text{ and } \triangle PAY) \quad \text{or} \quad \frac{BX}{PX} = \frac{AY}{PY}$$

This implies:

$$\frac{AY}{BX} = \frac{PY}{PX} \tag{III}$$

By multiplying (I), (II), and (III), we obtain:

$$\frac{BZ}{CY} \cdot \frac{CX}{AZ} \cdot \frac{AY}{BX} = \frac{PZ}{PY} \cdot \frac{PX}{PZ} \cdot \frac{PY}{PX} = 1 \text{ (or } -1, \text{ had we considered direction)}$$

Thus, by Menelaus's theorem, points X, Y, and Z are collinear. These three points determine the Simson line of $\triangle ABC$ with respect to point P. ●

Proof II (See Figure 3-21.) Because $\angle PYA$ is supplementary to $\angle PZA$, quadrilateral PZAY is cyclic. Draw $\overline{PA}$, $\overline{PB}$, and $\overline{PC}$. Therefore:

$$m\angle PYZ = m\angle PAZ \tag{I}$$

Similarly, because $\angle PYC$ is supplementary to $\angle PXC$, quadrilateral PXCY is cyclic, and therefore:

$$m\angle PYX = m\angle PCB \tag{II}$$

However, quadrilateral PACB is also cyclic because it is inscribed in the given circumcircle, and therefore:

$$m\angle PAZ(m\angle PAB) = m\angle PCB \tag{III}$$

From (I), (II), and (III), $m\angle PYZ = m\angle PYX$, and thus points X, Y, and Z are collinear. ●

For other proofs of Simson's theorem, see *Challenging Problems in Geometry* by A. S. Posamentier and C. T. Salkind (New York: Dover, 1996), pages 43–45.

There are many simple applications of the Simson line. We will consider a few of them.

Application 11 Sides $\overleftrightarrow{AB}$, $\overleftrightarrow{BC}$, and $\overleftrightarrow{CA}$ of $\triangle ABC$ are cut by a transversal at points Q, R, and S, respectively. The circumcircles of $\triangle ABC$ and $\triangle SCR$ intersect at point P. Prove that quadrilateral APSQ is cyclic. ●

Proof Draw perpendiculars $\overline{PX}$, $\overline{PY}$, $\overline{PZ}$, and $\overline{PW}$ to $\overleftrightarrow{AB}$, $\overleftrightarrow{AC}$, $\overleftrightarrow{QR}$, and $\overleftrightarrow{BC}$, respectively, as in Figure 3-22.

Because point P is on the circumcircle of $\triangle ABC$, points X, Y, and W are collinear (Simson's theorem). Similarly, because point P is on the circumcircle of $\triangle SCR$, points Y, Z, and W are collinear. It then follows that points X, Y, and Z are collinear.

Thus point P must lie on the circumcircle of $\triangle AQS$ (the converse of Simson's theorem, whose proof we leave as an exercise). Therefore quadrilateral APSQ is cyclic. ●

INTERACTIVE 3-22

Drag vertices *A, B,* and *C*
to change the shape of the
triangle and note that *APSQ* is
always cyclic.

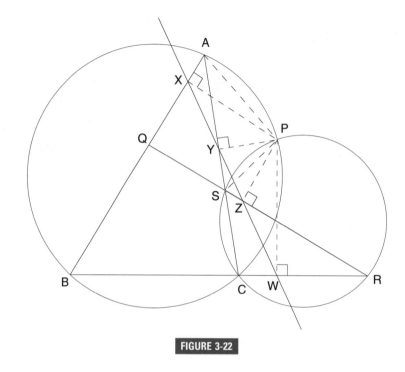

FIGURE 3-22

Application 12 $\overline{AB}$, $\overline{BC}$, $\overline{EC}$, and $\overline{ED}$ form triangles *ABC, FBD, EFA,* and *EDC.* Prove that the
four circumcircles of these triangles meet at a common point. ●

Proof   Consider the circumcircles of $\triangle ABC$ and $\triangle FBD$, which meet at points B and P.
From point P draw perpendiculars $\overline{PX}$, $\overline{PY}$, $\overline{PZ}$, and $\overline{PW}$ to $\overline{BC}$, $\overline{AB}$, $\overline{ED}$, and $\overline{EC}$,
respectively (see Figure 3-23). Because point P is on the circumcircle of $\triangle ABC$,
points X, Y, and W are collinear. Therefore points X, Y, Z, and W are collinear.

Because points Y, Z, and W are collinear, point P must lie on the circumcircle of $\triangle EFA$ (the converse of Simson's theorem). By the same reasoning, because
points X, Z, and W are collinear, point P lies on the circumcircle of $\triangle EDC$. Thus
all four circles pass through point P. ●

The Simson line has many interesting properties; we present a few here.

THEOREM 3.8   **(Simson line property I)**   If the altitude $\overline{AD}$ of $\triangle ABC$ meets the circumcircle at point P, then the Simson line of P with respect to $\triangle ABC$ is parallel to
the line tangent to the circle at point A.

INTERACTIVE 3-23

Drag vertices *A, B, C, D,* and *E* to
change the shape of the triangles
and see that the circles always
meet at a point.

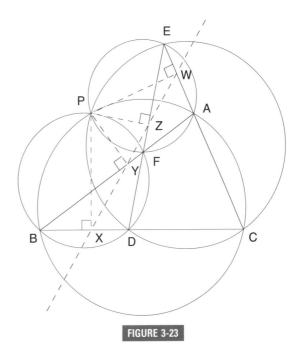

FIGURE 3-23

⋫roof   Because $\overline{PX}$ and $\overline{PZ}$ are perpendicular, respectively, to sides $\overleftrightarrow{AC}$ and $\overleftrightarrow{AB}$ of $\triangle ABC$,
points *X, D,* and *Z* determine the Simson line of point *P* with respect to $\triangle ABC$.

Draw $\overline{PB}$ (see Figure 3-24). Consider quadrilateral *PDBZ*, where $m\angle PDB = m\angle PZB = 90°$, thus making quadrilateral *PDBZ* a cyclic quadrilateral.* In
quadrilateral *PDBZ*:

$$m\angle DZB = m\angle DPB \qquad\qquad (I)$$

INTERACTIVE 3-24

Drag vertices *A, B,* and *C* to
change the shape of the triangle
and see that the required lines
are parallel.

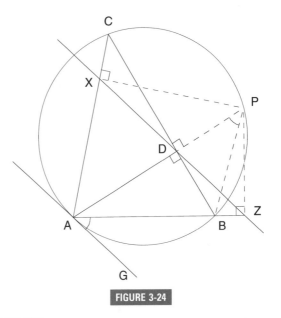

FIGURE 3-24

---

* Because opposite angles are supplementary.

However, in the circumcircle of $\triangle ABC$, $m\angle GAB = \frac{1}{2}(m\widehat{AB})$, and $m\angle DPB(m\angle APB) = \frac{1}{2}(m\widehat{AB})$. Therefore:

$$m\angle GAB = m\angle DPB \qquad (II)$$

From (I) and (II), by transitivity, $m\angle DZB = m\angle GAB$, and thus Simson line $\overleftrightarrow{XDZ}$ is parallel to tangent $\overleftrightarrow{GA}$. ●

THEOREM 3.9 **(Simson line property II)** From point P on the circumcircle of $\triangle ABC$, if perpendiculars $\overline{PX}$, $\overline{PY}$, and $\overline{PZ}$ are drawn to sides $\overleftrightarrow{AC}$, $\overleftrightarrow{AB}$, and $\overleftrightarrow{BC}$, respectively, then $(PA)(PZ) = (PB)(PX)$ (see Figure 3-25).

Proof Because $m\angle PYB = m\angle PZB = 90°$, quadrilateral $PYZB$ is cyclic.* Therefore:

$$m\angle PBY = m\angle PZY \qquad (I)$$

Likewise, because $m\angle PXA = m\angle PYA = 90°$, quadrilateral $PXAY$ is cyclic, and:

$$m\angle PXY = m\angle PAY \qquad (II)$$

Points X, Y, and Z are collinear (the Simson line). Therefore, from (I) and (II):

$$\triangle PAB \sim \triangle PXZ \Rightarrow \frac{PA}{PX} = \frac{PB}{PZ} \quad \text{or} \quad (PA)(PZ) = (PB)(PX) ●$$

INTERACTIVE 3-25

Drag vertices *A, B,* and *C* to change the shape of the triangle; drag point *P* on the circle and see that the products are equal.

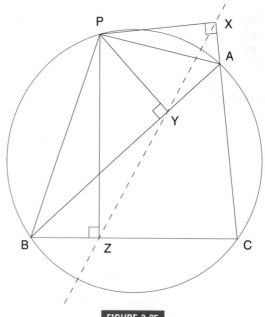

FIGURE 3-25

* A quadrilateral is cyclic (i.e., may be inscribed in a circle) if one side subtends congruent angles at the two opposite vertices.

THEOREM 3.10   **(Simson line property III)**   The measure of the angle determined by the Simson line of two given points on the circumcircle of a given triangle is equal to one-half the measure of the arc determined by the two points.

Proof   In Figure 3-26, $\overleftrightarrow{XYZ}$ is the Simson line for point P and $\overleftrightarrow{UVW}$ is the Simson line for point Q. Extend $\overline{PX}$ and $\overline{QW}$ to meet the circle at points M and N, respectively. Then draw $\overline{AM}$ and $\overline{AN}$. Because $m\angle PZB = m\angle PXB = 90°$, quadrilateral $PZXB$ is cyclic, and:

$$m\angle ZXP = m\angle ZBP \qquad\qquad \text{(I)}$$
$$m\angle ABP = m\angle AMP \quad \text{or} \quad m\angle ZBP = m\angle AMP \qquad \text{(II)}$$

From (I) and (II), $m\angle ZXP = m\angle AMP$. Therefore:

$$\overleftrightarrow{XYZ} \parallel \overline{AM}$$

In a similar fashion, it may be shown that $\overleftrightarrow{UVW} \parallel \overline{AN}$.

Hence, if T is the point of intersection of the two Simson lines, then $m\angle XTW = m\angle MAN$ because their corresponding sides are parallel. Now, $m\angle MAN = \frac{1}{2}(m\widehat{MN})$, but because $\overline{PM} \parallel \overline{QN}$, we have $m\widehat{MN} = m\widehat{PQ}$ and therefore $m\angle MAN = \frac{1}{2}(m\widehat{PQ})$. Thus $m\angle XTW = \frac{1}{2}(m\widehat{PQ})$. ●

INTERACTIVE 3-26

Drag vertices *A*, *B*, and *C* to change the shape of the triangle; drag points *P* and *Q* on the circle, and the angle is one-half the measure of the arc.

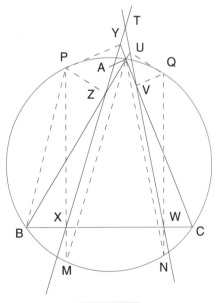

FIGURE 3-26

Here is an interesting application of Simson's theorem to an earlier dilemma. Recall the proof of the fallacy that all scalene triangles are isosceles in Chapter 1. Perpendiculars were drawn from point G to $\overleftrightarrow{AC}$, $\overleftrightarrow{BC}$, and $\overleftrightarrow{AB}$, meeting these lines at points D, F, and E, respectively. Because point G is on the circumcircle of $\triangle ABC$, Simson's theorem establishes that points D, E, and

F are collinear. The famous postulate by Moritz Pasch (1843–1930) states that a straight line intersecting one side of a triangle (internally) must intersect exactly one of the other two sides (internally) except if the line contains a vertex of the triangle. Euclid had quietly assumed this idea. Yet with this postulate available to us, we are assured that the two critical perpendiculars can neither both fall inside nor both fall outside the triangle, which enables us to avoid the fallacious proof previously offered.

RADICAL AXES

Earlier in this chapter, we stated Brianchon's theorem as the dual of Pascal's theorem. At that juncture we deferred the proof because we needed some more knowledge about a radical axis. We will now establish some important properties of radical axes and then use them to prove Brianchon's theorem.

Consider two circles *R* and *Q* (see Figure 3-27) intersecting at points *A* and *B*. *P* is any point on $\overleftrightarrow{AB}$ not between points *A* and *B*. $\overline{PT}$ and $\overline{PS}$ are tangents to circles *R* and *Q* at points *T* and *S*, respectively.

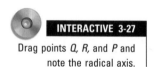

INTERACTIVE 3-27

Drag points *Q, R,* and *P* and note the radical axis.

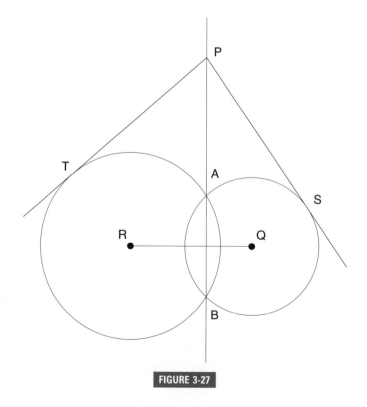

FIGURE 3-27

From our study of elementary geometry, we know that PT is the mean proportional between PB and PA. Therefore $(PT)^2 = (PB)(PA)$. Similarly, for circle Q, $(PS)^2 = (PB)(PA)$. It then follows that $PT = PS$.

Because point P was selected as *any* external point on $\overleftrightarrow{AB}$, we can conclude that from any external point on $\overleftrightarrow{AB}$, tangent segments to circles R and Q are congruent. Before we can state this as a locus theorem, we must prove that any point P that generates congruent tangents to circles R and Q must lie on $\overleftrightarrow{AB}$.

Suppose P is any point where tangent segments $\overline{PT}$ and $\overline{PS}$ are congruent. Let $\overrightarrow{PA}$ intersect circle R at point B and circle Q at point B'. As before, $(PB)(PA) = (PT)^2$ and $(PB')(PA) = (PS)^2$. Because $PT = PS$, $PB = PB'$. Therefore B and B' must coincide, and P lies on the common secant, $\overleftrightarrow{PA}$, of the two circles. We call the line consisting of points that are common endpoints of congruent tangent segments to two circles the *radical axis* of the two circles.

We now state this result as our next theorem.

THEOREM 3.11   The radical axis of two intersecting circles is their common secant.

It follows immediately that the radical axis of two tangent circles is their common tangent. Before we can investigate the radical axis of two nonintersecting circles, we need to consider the following theorem.

THEOREM 3.12   The locus of a point the difference of whose distances squared from two fixed points is a constant is a line perpendicular to the segment determined by the two fixed points.

Proof   Let R and Q be the fixed points and let P be a point on the locus (see Figure 3-28). Draw $\overline{PR}$ and $\overline{PQ}$. Construct $\overline{PN} \perp \overline{RQ}$. We use the Pythagorean theorem to get:

$$(PR)^2 - (RN)^2 = (PN)^2 \quad \text{and} \quad (PQ)^2 - (QN)^2 = (PN)^2$$

Therefore:

$$(PR)^2 - (RN)^2 = (PQ)^2 - (QN)^2 \quad \text{or} \quad (PR)^2 - (PQ)^2 = (RN)^2 - (QN)^2 = k$$

Let $RQ = d$. Then, by factoring the last equality, we have:

$$(RN + QN)(RN - QN) = k$$

$$d(RN - QN) = k$$

$$RN - QN = \frac{k}{d} \tag{I}$$

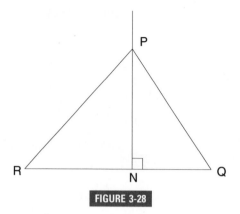

FIGURE 3-28

Remember that:

$$RN + QN = d \qquad \text{(II)}$$

Solving equations (I) and (II) simultaneously, we get:

$$RN = \frac{d^2 + k}{2d} \quad \text{and} \quad QN = \frac{d^2 - k}{2d}$$

This fixes the position of point N.

Because d and k are constant for any given situation, point P must lie on the line perpendicular to $\overline{RQ}$ at point N, which divides $\overline{RQ}$ in the ratio:

$$\frac{RN}{QN} = \frac{d^2 + k}{d^2 - k}$$

We can conclude this locus proof by showing that any point on $\overleftrightarrow{PN}$ satisfies the given conditions. This is left to the reader. ●

Theorem 3.12 enables us to continue our study of radical axes. We must now determine the radical axis of two nonintersecting circles. Our intuition would probably predict the next theorem.

THEOREM 3.13   The radical axis of two nonintersecting circles is a line perpendicular to their line of centers.

℗roof   Begin by letting r and q be the radii of circles R and Q, respectively. Let P be a point on the required locus, that is, so that tangent segments $\overline{PT}$ and $\overline{PS}$ are congruent (see Figure 3-29).

By applying the Pythagorean theorem to $\triangle PTR$ and $\triangle PSQ$, we get:

$$(PR)^2 - r^2 = (PT)^2 \quad \text{and} \quad (PQ)^2 - q^2 = (PS)^2$$

But $PT = PS$; therefore:

$$(PR)^2 - r^2 = (PQ)^2 - q^2 \quad \text{or} \quad (PR)^2 - (PQ)^2 = r^2 - q^2$$

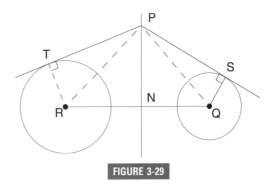

FIGURE 3-29

Because the right-hand side of this equality is a constant, we can conclude (by Theorem 3.12) that the locus of P is the line containing point P, which is perpendicular to the line of centers $\overleftrightarrow{RQ}$. ●

In a manner similar to that used in the previous proof, we can determine the location of point N in terms of the radii and the distance between the centers. As a direct consequence of Theorem 3.13, we have the following theorem.

THEOREM 3.14 The radical axes of three given circles whose centers are not collinear are concurrent.

℗roof Let us consider circles R, Q, and U whose radical axes are $\overleftrightarrow{AB}$, $\overleftrightarrow{CD}$, and $\overleftrightarrow{EF}$ (see Figure 3-30).

Let point P be the intersection of $\overleftrightarrow{AB}$ and $\overleftrightarrow{CD}$. Using radical axis $\overleftrightarrow{AB}$ of circles R and Q, we have $PT = PS$. Using radical axis $\overleftrightarrow{CD}$ of circles Q and U, we have $PV = PS$. (Note: $\overleftrightarrow{PT}$, $\overleftrightarrow{PS}$, and $\overleftrightarrow{PV}$ are tangents to the given circles.)

Thus $PT = PV$, which indicates that point P must lie on the radical axis, $\overleftrightarrow{EF}$, of circles R and U. This proves that the radical axes are concurrent at point P. ●

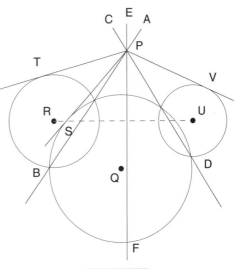

FIGURE 3-30

We are now ready to prove Brianchon's theorem, which we discussed earlier in the chapter (on page 58). The proof we will use is by A. S. Smogorzhevskii (*The Ruler in Geometrical Constructions,* New York: Blaisdell Publishing Company, 1961, pp. 33–35).

THEOREM 3.5   **(Brianchon's theorem)**   If a hexagon is circumscribed about a circle, the lines containing opposite vertices are concurrent.

℗roof   As seen in Figure 3-31, the sides of hexagon *ABCDEF* are tangent to a circle at points *T, N, L, S, M,* and *K.* Points *K′, L′, N′, M′, S′,* and *T′* are chosen on $\overrightarrow{FA}$, $\overrightarrow{DC}$, $\overrightarrow{BC}$, $\overrightarrow{FE}$, $\overrightarrow{DE}$, and $\overrightarrow{BA}$, respectively, so that:

$$KK' = LL' = NN' = MM' = SS' = TT'$$

Now construct circle *P* tangent to $\overrightarrow{BA}$ and $\overrightarrow{DE}$ at points *T′* and *S′*, respectively (the existence of this circle is easily justified). Similarly, construct circle *Q* tangent to $\overrightarrow{FA}$ and $\overrightarrow{DC}$ at points *K′* and *L′*, respectively. Then construct circle *R* tangent to $\overrightarrow{FE}$ and $\overrightarrow{BC}$ at points *M′* and *N′*, respectively.

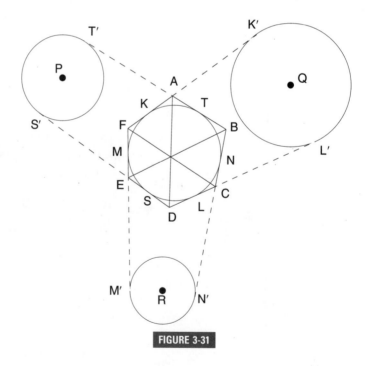

FIGURE 3-31

Because two tangent segments to a circle from an external point have the same length, $FM = FK$. We already know that $MM' = KK'$. Therefore, by addition:

$$FM' = FK'$$

Similarly:

$$CL = CN \quad \text{and} \quad LL' = NN'$$

By subtraction:

$$CL' = CN'$$

We now notice that points F and C are each endpoints of a pair of congruent tangent segments to circles R and Q. Thus these points determine the radical axis, $\overleftrightarrow{CF}$, of circles R and Q. Using the same technique, we can easily show that $\overleftrightarrow{AD}$ is the radical axis of circles P and Q and that $\overleftrightarrow{BE}$ is the radical axis of circles P and R.

We proved that the radical axes of three circles with noncollinear centers (taken in pairs) are concurrent (Theorem 3.14). Therefore $\overleftrightarrow{CF}$, $\overleftrightarrow{AD}$, and $\overleftrightarrow{BE}$ are concurrent.

We should note that the only way in which these circles would have had collinear centers is if the diagonals were to have coincided, which is impossible.

———————————————— E X E R C I S E S ————————————————

1. Sides $\overleftrightarrow{AB}$, $\overleftrightarrow{BC}$, $\overleftrightarrow{CD}$, and $\overleftrightarrow{DA}$ of quadrilateral $ABCD$ are intersected by a straight line at points K, L, M, and N, respectively. Prove that
$$\frac{BL}{LC} \cdot \frac{AK}{KB} \cdot \frac{DN}{NA} \cdot \frac{CM}{MD} = 1.$$

2. Side $\overline{AB}$ of square $ABCD$ is extended to point P so that $BP = 2(AB)$. With M the midpoint of $\overline{DC}$, $\overline{BM}$ intersects $\overline{AC}$ at point Q. Also, $\overline{PQ}$ intersects $\overline{BC}$ at point R. Use Menelaus's theorem to find the numerical value of $\dfrac{CR}{RB}$.

3. Points P and R are on sides $\overline{AB}$ and $\overline{AC}$, respectively, of $\triangle ABC$ so that $\overline{AP} \cong \overline{AR}$. Prove that median $\overline{AM}$ partitions $\overline{PR}$ into segments proportional to $\overline{AB}$ and $\overline{AC}$.

4. Prove that the tangents to the circumcircle of a triangle at its vertices intersect the opposite sides in three collinear points.

5. Prove that if a line contains the centroid, G, of $\triangle ABC$ and intersects sides $\overleftrightarrow{AB}$ and $\overleftrightarrow{AC}$ at points M and N, respectively, then $(AM)(NC) + (AN)(MB) = (AM)(AN)$.

6. A circle is tangent to side $\overline{BC}$ of $\triangle ABC$ at M, its midpoint, and intersects $\overleftrightarrow{AB}$ and $\overleftrightarrow{AC}$ at points R, R' and S, S', respectively. If $\overline{RS}$ and $\overline{R'S'}$ are each extended to meet $\overleftrightarrow{BC}$ at points P and P', respectively, prove that $(BP)(BP') = (CP)(CP')$ (see Figure 3-32).

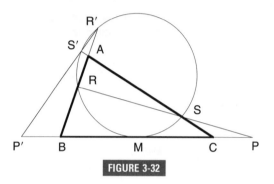

FIGURE 3-32

7. In $\triangle ABC$, P, Q, and R are the midpoints of sides $\overline{AB}$, $\overline{BC}$, and, $\overline{AC}$ respectively. $\overleftrightarrow{AN}$, $\overleftrightarrow{BL}$, and $\overleftrightarrow{CM}$ are concurrent, meeting the opposite sides at points N, L, and M, respectively. If $\overleftrightarrow{PL}$ meets $\overleftrightarrow{BC}$ at point J, $\overleftrightarrow{MQ}$ meets $\overleftrightarrow{AC}$ at point I, and $\overleftrightarrow{RN}$ meets $\overleftrightarrow{AB}$ at point H, prove that points H, I, and J are collinear (see Figure 3-33).

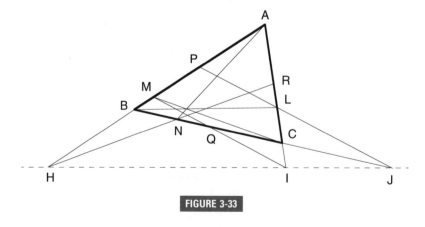

FIGURE 3-33

8. Prove that the three pairs of common external tangents to three circles (no two of which are equal or concentric), taken two at a time, intersect in three collinear points.

9. Prove that the perpendicular bisectors of the interior angle bisectors of any triangle meet the sides opposite the angles being bisected in three collinear points.

10. Provide a proof for Application 6, using Menelaus's theorem.

11. How can Brianchon's theorem be used to prove the existence of the Gergonne point of a triangle?

12. Compare Pappus's theorem to Pascal's theorem.

13. State and prove the converse of Desargues's theorem.

14. State and prove the converse of Simson's theorem.

15. In Figure 3-34, $\triangle ABC$, with right angle at A, is inscribed in circle O. The Simson line of point P with respect to $\triangle ABC$ meets $\overline{PA}$ at point M. Prove that $\overline{MO}$ is perpendicular to $\overline{PA}$ at point M.

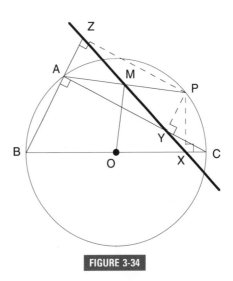

FIGURE 3-34

16. From a point P on the circumference of circle O, three chords are drawn meeting the circle at points A, B, and C. Prove that the three points of intersection of the three circles with $\overline{PA}$, $\overline{PB}$, and $\overline{PC}$ as diameters are collinear.

17. If two triangles are inscribed in the same circle, a single point on the circumcircle determines a Simson line for each triangle. Prove that the angle formed by these two Simson lines is constant, regardless of the position of the point.

18. Prove that the common tangent segments (if they exist) of two given circles are bisected by their radical axis.

19. Prove that the radical axis of the two circles whose diameters are the diagonals of a trapezoid contains the point of intersection of the nonparallel sides of the trapezoid.

20. Prove that the four points determined by the intersections of two secants drawn from a point on the radical axis of two circles with the two circles lie on a third circle.

SOME
SYMMETRIC
POINTS IN A
TRIANGLE

Suppose that you and one of your friends are planning to set up a special remote computer server to store the data from your computers and that at your school. The three locations determine a triangle with no angle greater than 120°. Using a map of your town (see Figure 4.0), you seek a location for this computer that makes the sum of the distances from the computer to each of your houses a minimum. We will call this point the "minimum distance point." How would you find this point?

In this chapter, we will develop some theorems that will enable us to solve this problem. Along the way, we will encounter a number of interesting theorems that highlight some fascinating properties of triangles.

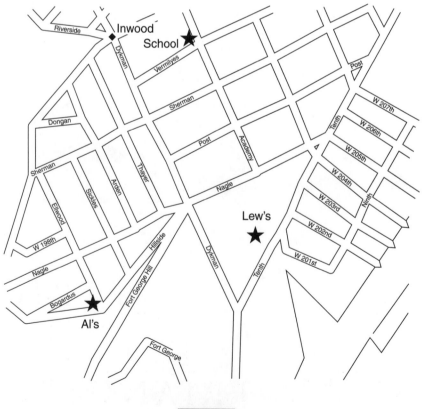

FIGURE 4-0

EQUIANGULAR POINT

INTERACTIVE 4-1

Drag points *A*, *B*, and *C* to change the shape of the triangle and see that *O* remains the equiangular point.

Consider any convenient triangle. How would you locate the point in the triangle at which congruent angles are formed by drawing rays from this point to the vertices?

Let us set out to locate this point (see Figure 4-1). We will first find a point that has another interesting property. Begin by constructing an equilateral triangle externally on each side of the given triangle. Draw segments joining each vertex of the given triangle with the remote vertex of the equilateral triangle on the opposite side (see Figure 4-2). Theorem 4.1 presents an astonishing property of these three line segments. After proving this property, we will return to our original problem.

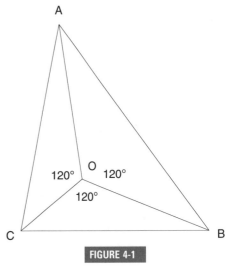

FIGURE 4-1

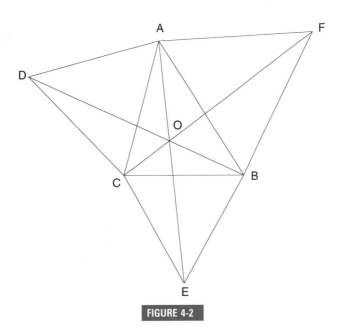

FIGURE 4-2

█ THEOREM 4.1   The segments joining each vertex of a given triangle with the remote vertex of the equilateral triangle (drawn externally on the opposite side of the given triangle) are congruent.

Plan of Proof   Prove $\overline{DB} \cong \overline{AE}$ and $\overline{AE} \cong \overline{CF}$, by first proving $\triangle DCB \cong \triangle ACE$ and then proving $\triangle EBA \cong \triangle CBF$. ●

Proof   Because $m\angle DCA = m\angle ECB = 60°$, $m\angle DCB = m\angle ACE$ (by addition). Also, because we have equilateral triangles, $DC = AC$ and $CB = CE$. Therefore $\triangle DCB \cong \triangle ACE$ (SAS) and $\overline{DB} \cong \overline{AE}$. In a similar manner, we can prove that $\triangle EBA \cong \triangle CBF$. This enables us to conclude that $\overline{AE} \cong \overline{CF}$. Thus $\overline{DB} \cong \overline{AE} \cong \overline{CF}$. ●

From the diagram in Figure 4-2, it appears that $\overline{DB}$, $\overline{AE}$, and $\overline{CF}$ are concurrent. This observation gives us our next theorem.

▌THEOREM 4.2   The segments joining each vertex of a given triangle with the remote vertex of the equilateral triangle drawn externally on the opposite side of the given triangle are concurrent. (This point is called the *Fermat point** of the triangle.)

Plan of Proof   Construct the circumcircle of each of the three equilateral triangles. Then show that the three circles are concurrent at point O. The six segments from point O to points A, B, C, D, E, and F will determine the three concurrent lines. ●

Proof   Consider the circumcircles of the three equilateral triangles ACD, ABF, and BCE. Let K, L, and M be the centers of these circles (see Figure 4-3).

Circles K and L meet at points O and A. Because $m\overset{\frown}{ADC} = 240°$ and because we know that $m\angle AOC = \frac{1}{2}(m\overset{\frown}{ADC})$, $m\angle AOC = 120°$. Similarly, $m\angle AOB = \frac{1}{2}(m\overset{\frown}{AFB}) = 120°$. Therefore $m\angle COB = 120°$ (because a complete revolution $= 360°$).

Because $m\overset{\frown}{CEB} = 240°$, $\angle COB$ is an inscribed angle and point O must lie on circle M. Therefore we see that the three circles are concurrent, intersecting at point O.

Now join point O with points A, B, C, D, E, and F. $m\angle DOA = m\angle AOF = m\angle FOB = 60°$, and therefore $\overleftrightarrow{DOB}$. Similarly, $\overleftrightarrow{COF}$ and $\overleftrightarrow{AOE}$. Thus it has been proved that $\overline{DB}$, $\overline{AE}$, and $\overline{CF}$ are concurrent, intersecting at point O (which is also the point of intersection of circles K, L, and M). ●

Can you now determine the point in $\triangle ABC$ at which the three sides subtend (i.e., determine by being opposite) congruent angles? The point O is called the *equiangular point* of $\triangle ABC$ because $m\angle AOB = m\angle AOC = m\angle BOC = 120°$. We will be referring to this point again later in the chapter.

Before continuing with our search for the equiangular point, let us take advantage of another interesting property. Sources indicate that the following theorem was developed by Napoleon Bonaparte, who took pride in his mathematical talents. Thus the resulting equilateral triangle is often called the *Napoleon triangle*.

---

* Named after French mathematician Pierre de Fermat (1601–1665).

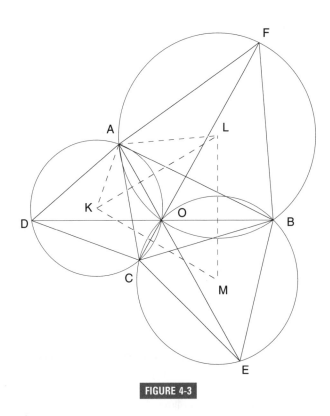

FIGURE 4-3

| **THEOREM 4.3** | The circumcenters of the three equilateral triangles drawn externally on the sides of a given triangle determine an equilateral triangle. |

Plan of Proof   Prove that the sides of $\triangle KLM$ are proportional to $\overline{AE}$, $\overline{BD}$, and $\overline{CF}$. (We have previously proved that $\overline{DB} \cong \overline{AE} \cong \overline{CF}$.) ●

Proof   Consider $\triangle DAC$ (see Figure 4-3). Because K is the centroid (the point of intersection of the medians) of $\triangle DAC$, AK is two-thirds the length of the altitude (or median). Using the relationships in a 30-60-90 triangle, we find that $AC{:}AK = \sqrt{3}{:}1$. Similarly, in equilateral triangle AFB, $AF{:}AL = \sqrt{3}{:}1$. Therefore $AC{:}AK = AF{:}AL$.

Because $m\angle KAC = m\angle LAF = 30°$, $m\angle CAL = m\angle CAL$ (reflexive), and $m\angle KAL = m\angle CAF$ (addition), we have $\triangle KAL \sim \triangle CAF$. Thus $CF{:}KL = CA{:}AK = \sqrt{3}{:}1$.

Similarly, we can prove that $DB{:}KM = \sqrt{3}{:}1$ and $AE{:}ML = \sqrt{3}{:}1$. Therefore $DB{:}KM = AE{:}ML = CF{:}KL$. But because $DB = AE = CF$, as proved earlier, we obtain $KM = ML = KL$. Therefore $\triangle KML$ is equilateral. ●

A PROPERTY OF EQUILATERAL TRIANGLES

We need to develop one more surprising fact about equilateral triangles before we consider our initial problem involving the remote computer server.

Draw a large equilateral triangle. Choose any convenient point in the interior region of this triangle. Now measure the distances from this point to the three sides and record the sum of these distances. (This can be done either on paper or on The Geometer's Sketchpad®.)

Repeat this procedure for any other point in the interior region of this triangle. How do the two sums compare? Now measure the length of the altitude of the triangle. How do the two sums compare to the length of the altitude of the equilateral triangle? The answers to these questions suggest the following theorem.

THEOREM 4.4
The sum of the distances from any point in the interior of an equilateral triangle to the sides of the triangle is constant (the length of the altitude of the triangle).

We provide two proofs of this interesting property here. The first compares the length of each perpendicular segment to a portion of the altitude, and the second involves area comparisons.

Proof I
In equilateral triangle ABC, $\overline{PR} \perp \overline{AC}$, $\overline{PQ} \perp \overline{BC}$, $\overline{PS} \perp \overline{AB}$, and $\overline{AD} \perp \overline{BC}$. Draw a line through point P parallel to $\overline{BC}$, meeting $\overline{AD}$, $\overline{AB}$, and $\overline{AC}$ at points G, E, and F, respectively (see Figure 4-4).

Because quadrilateral $PGDQ$ is a rectangle, $PQ = GD$. Draw $\overline{ET} \perp \overline{AC}$. Because $\triangle AEF$ is equilateral, $\overline{AG} \cong \overline{ET}$ (all the altitudes of an equilateral triangle are congruent). Draw $\overline{PH} \parallel \overline{AC}$, meeting $\overline{ET}$ at point N. $\overline{NT} \cong \overline{PR}$. Because

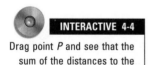

INTERACTIVE 4-4

Drag point P and see that the sum of the distances to the sides is constant.

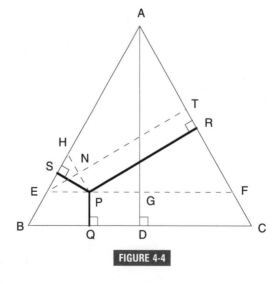

FIGURE 4-4

$\triangle EHP$ is equilateral, altitudes $\overline{PS}$ and $\overline{EN}$ are congruent. Therefore we have shown that $PS + PR = ET = AG$. Because $PQ = GD$:

$$PS + PR + PQ = AG + GD = AD \text{ (a constant for the given triangle)} \bullet$$

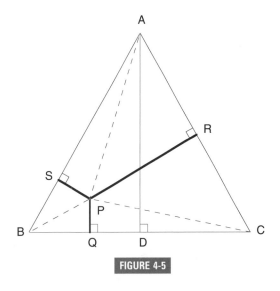

FIGURE 4-5

Proof II   In equilateral triangle ABC, $\overline{PR} \perp \overline{AC}$, $\overline{PQ} \perp \overline{BC}$, $\overline{PS} \perp \overline{AB}$, and $\overline{AD} \perp \overline{BC}$. Draw $\overline{PA}$, $\overline{PB}$, and $\overline{PC}$ (see Figure 4-5).

$$\text{area } \triangle ABC = \text{area } \triangle APB + \text{area } \triangle BPC + \text{area } \triangle CPA$$
$$= \frac{1}{2}(AB)(PS) + \frac{1}{2}(BC)(PQ) + \frac{1}{2}(AC)(PR)$$

Because $AB = BC = AC$, area $\triangle ABC = \frac{1}{2}(BC)[PS + PQ + PR]$. However, area $\triangle ABC = \frac{1}{2}(BC)(AD)$. Therefore:

$$PS + PQ + PR = AD \text{ (a constant for the given triangle)} \bullet$$

A MINIMUM DISTANCE POINT

Before we tackle our original problem of finding the minimum distance point of a triangle, let us consider a quadrilateral. For which point in a quadrilateral do you think the sum of the distances to the vertices would be less than that for any other point (i.e., a minimum sum)? Your first guess was probably correct—the point of intersection of the diagonals, which we call the minimum distance point of a quadrilateral. Now let us verify this guess.

To prove that among the interior points of a quadrilateral the diagonal-intersection point has the smallest sum of distances to the vertices, we simply choose any other interior point and compare its sum of distances to the vertices to that of the diagonal-intersection point.

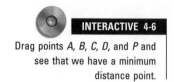

Drag points *A, B, C, D,* and *P* and see that we have a minimum distance point.

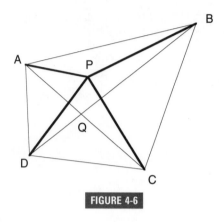

FIGURE 4-6

Consider quadrilateral *ABCD* with diagonals $\overline{AC}$ and $\overline{BD}$ intersecting at point *Q*. Select any point *P* (not at *Q*) in the interior of quadrilateral *ABCD* (see Figure 4-6). $PA + PC > QA + QC$ (because the sum of the lengths of two sides of a triangle is greater than the length of the third). Similarly, $PB + PD > QB + QD$. By addition, $PA + PB + PC + PD > QA + QB + QC + QD$, which shows that the sum of the distances from the point of intersection of the diagonals of a quadrilateral to the vertices is less than the sum of the distances from *any other* interior point of the quadrilateral to the vertices. This allows us to state the following theorem.

THEOREM 4.5   The minimum distance point of a quadrilateral is the point of intersection of the diagonals.

It is quite natural to wonder where the minimum distance point of a triangle would be. This is precisely the problem posed at the beginning of this chapter. As you ponder this problem, you are probably seeking a symmetric point in a given triangle. Perhaps you consider the equiangular point, certainly a point that offers some symmetry. Let us build on this guess.

Consider $\triangle ABC$ with no angle measuring greater than 120°. Let *M* be the point in the interior of $\triangle ABC$, where $m\angle AMB = m\angle BMC = m\angle AMC = 120°$ (see Figure 4-7). Draw lines through *A, B,* and *C* that are perpendicular to $\overline{AM}, \overline{BM},$ and $\overline{CM}$, respectively. These lines meet to form equilateral triangle *PQR*. (To prove that $\triangle PQR$ is equilateral, notice that each angle has measure 60°. This can be shown by considering, for example, quadrilateral *AMBR*. Because $m\angle RAM = m\angle RBM = 90°$ and $m\angle AMB = 120°$, it follows that $m\angle ARB = 60°$.)

Let *D* be *any other* point in the interior of $\triangle ABC$. We must show that the sum of the distances from point *M* to the vertices is less than the sum of the distances from point *D* to the vertices. From Theorem 4.4, we know that $MA + MB + MC = DE + DF + DG$ (where $\overline{DE}, \overline{DF},$ and $\overline{DG}$ are the perpendiculars to $\overline{REQ}, \overline{RBP},$ and $\overline{QGP}$, respectively). But $DE + DF + DG < DA + DB + DC$. (The shortest distance from an external point to a line is the length of the perpendicular segment from the point to the line.) By substitution:

$$MA + MB + MC < DA + DB + DC$$

INTERACTIVE 4-7

Drag points *A*, *B*, *C*, and *D* and see that we have a minimum distance point.

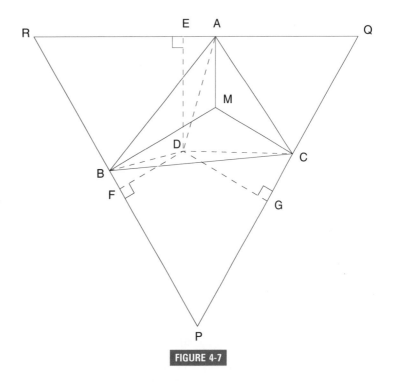

FIGURE 4-7

You may wonder why we chose to restrict our discussion to triangles with angles of measure less than 120°. If you try to construct the point *M* in a triangle with one angle of measure of 150°, the reason for our restriction will become obvious.

THEOREM 4.6   The minimum distance point of a triangle (with no angle of measure greater than 120°) is the equiangular point (i.e., the point at which the sides of the triangle subtend congruent angles).

We are now prepared to solve the original problem involving the best location of the remote computer (i.e., the location where the sum of the distances to the three houses is a minimum). After drawing a triangle on the map (with the three houses as vertices), you would construct the minimum distance point by constructing the equiangular point (which is also the minimum distance point) in the manner described for Theorem 4.1.

─────────────── E X E R C I S E S ───────────────

1. Find the sum of the lengths of the three perpendicular segments from any point in an equilateral triangle to each of the sides, each of which has length 10.

2. Locate by construction the point in a given acute triangle that has the least sum of distances to the vertices.

3. Explain why the 120° restriction is placed on Theorem 4.6.

4. If one angle of a triangle has measure greater than or equal to 120°, prove that the vertex of this angle is the minimum distance point of the triangle.

5. If squares are constructed externally on the sides of a triangle, prove that the line containing the centers of any two of these squares is perpendicular to the line containing the common vertex of these two squares and the center of the third square.

6. Prove that of all triangles with a given perimeter, the one with the greatest area is the equilateral triangle.

7. Prove that of all triangles with a given area, the one with the least perimeter is the equilateral triangle.

8. Prove that if similar triangles are erected externally on the sides of any triangle, the triangle formed by the circumcenters of the three similar triangles determines a triangle similar to the three triangles.

9. Prove Theorem 4.3 for the case in which the three equilateral triangles are drawn internally. (This is called the *internal Napoleon triangle,* whereas the triangle for Theorem 4.3 is called the *external Napoleon triangle.*)

10. Prove that the external Napoleon triangle and the internal Napoleon triangle have the same center and that their areas differ by the area of the original triangle.

CHAPTER
FIVE

MORE
TRIANGLE
PROPERTIES

It is generally accepted (especially by high school geometry students) that the study of triangle properties forms the foundation for the study of synthetic geometry. After completing the high school geometry course, students tend to feel that they know all there is to know about triangles. Having reached this point in the book, you can clearly see that this is not so. However, you may still feel that within the realm of "elementary geometry" your knowledge about triangles is complete. This may very well be the case. Read on and see how some seemingly innocent properties of triangles are, in fact, not so trivial after all.

ANGLE BISECTORS

Early in their studies, all high school geometry students learn that *the angle bisectors of the base angles of an isosceles triangle are congruent.* This is rather easily proved. Yet the converse of this statement is conspicuously omitted. It, too, is a valid theorem but is quite difficult to prove.

THEOREM 5.1	If two angle bisectors of a triangle are congruent, then the triangle is isosceles.

The proof of this theorem is regarded as one of the most difficult in elementary geometry. For this reason, we provide a number of different proofs of this theorem here. Each is instructional and merits special attention. We first restate the theorem for $\triangle ABC$.

GIVEN: $\overline{AE}$ and $\overline{BD}$ are angle bisectors of $\triangle ABC$. $\overline{AE} \cong \overline{BD}$ (see Figure 5-1).

PROVE: $\triangle ABC$ is isosceles.

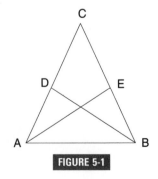

FIGURE 5-1

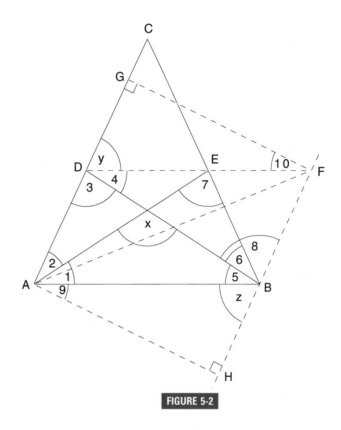

FIGURE 5-2

Ⓟroof I Draw $\angle DBF \cong \angle AEB$ so that $\overline{BF} \cong \overline{BE}$. Then draw $\overline{DF}$ (see Figure 5-2). Also draw $\overline{FG} \perp \overline{AC}$ and $\overline{AH} \perp \overline{FB}$ at H.

Because by hypothesis $\overline{AE} \cong \overline{DB}$, $\overline{FB} \cong \overline{EB}$, and $\angle 8 \cong \angle 7$, it follows that $\triangle AEB \cong \triangle DBF$ (SAS), so $DF = AB$ and $m\angle 1 = m\angle 4$.

$$m\angle x = m\angle 2 + m\angle 3 \quad \text{(exterior angle of a triangle)}$$
$$m\angle x = m\angle 1 + m\angle 3 \quad \text{(substitution)}$$
$$m\angle x = m\angle 4 + m\angle 3 \quad \text{(substitution)}$$
$$m\angle x = m\angle 7 + m\angle 6 \quad \text{(exterior angle of a triangle)}$$
$$m\angle x = m\angle 7 + m\angle 5 \quad \text{(substitution)}$$
$$m\angle x = m\angle 8 + m\angle 5 \quad \text{(substitution)}$$

Therefore:

$$m\angle 4 + m\angle 3 = m\angle 8 + m\angle 5 \quad \text{(transitivity)}$$

Thus:

$$m\angle z = m\angle y$$

Right triangle $FDG \cong$ right triangle ABH (SAA), $DG = BH$, and $FG = AH$. Right triangle $AFG \cong$ right triangle FAH (HL), and $AG = FH$.

Therefore quadrilateral *GFHA* is a parallelogram. Also, $m\angle 9 = m\angle 10$ (from $\triangle ABH$ and $\triangle FDG$).

$$m\angle DAB = m\angle DFB \quad \text{(subtraction)}$$
$$m\angle DFB = m\angle EBA \quad \text{(from } \triangle DBF \text{ and } \triangle AEB)$$

Therefore $m\angle DAB = m\angle EBA$ (by transitivity), and $\triangle ABC$ is isosceles. ●

Proof II (indirect) Assume $\triangle ABC$ is *not* isosceles. Let $m\angle ABC > m\angle ACB$ (see Figure 5-3). By hypothesis, $\overline{BF} \cong \overline{CE}$. $\overline{BC} \cong \overline{BC}$ and $\overline{CF} > \overline{BE}$.* Through point *F*, construct $\overline{GF}$ parallel to $\overline{EB}$. Through point *E*, construct $\overline{GE}$ parallel to $\overline{BF}$. Therefore quadrilateral *BFGE* is a parallelogram, $\overline{BF} \cong \overline{EG}$, $\overline{EG} \cong \overline{CE}$, and $\triangle GEC$ is isosceles.

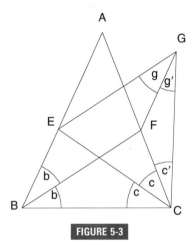

FIGURE 5-3

$$m\angle(g + g') = m\angle(c + c') \quad \text{and} \quad m\angle g = m\angle b$$

Thus:

$$m\angle(b + g') = m\angle(c + c')$$

Because $m\angle b > m\angle c$ (by hypothesis):

$$m\angle g' < m\angle c'$$

In $\triangle GFC$, we have $CF < GF$. But $GF = BE$. Thus $CF < BE$. The assumption of the inequality of $m\angle ABC$ and $m\angle ACB$ leads to two contradictory results:

$$CF < BE \quad \text{and} \quad CF > BE$$

Therefore $\triangle ABC$ is isosceles. ●

_____

* If two triangles have two pairs of corresponding sides congruent and their included angles are not congruent, then the greater third side is opposite the greater included angle.

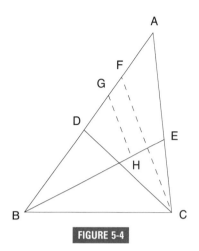

FIGURE 5-4

℗roof III (indirect) In △ABC (see Figure 5-4), the bisectors of angles ABC and ACB have equal lengths (i.e., BE = DC).

Assume that m∠ABC < m∠ACB. We then draw ∠FCD congruent to ∠ABE. Note that we may take point F between vertices B and A without loss of generality. In △FBC, FB > FC. (If the measures of two angles of a triangle are not equal, then the measures of the sides opposite these angles are also unequal, the side with the greater measure being opposite the angle with the greater measure.) Choose a point G so that $\overline{BG} \cong \overline{FC}$. Then draw $\overline{GH} \parallel \overline{FC}$.

Therefore ∠BGH ≅ ∠BFC (corresponding angles) and △BGH ≅ △CFD (ASA). It follows that BH = DC. Because BH < BE, this contradicts the hypothesis that the lengths of the angle bisectors $\overline{DC}$ and $\overline{BE}$ are equal. A similar argument will show that it is impossible to have m∠ACB < m∠ABC.

It then follows that m∠ACB = m∠ABC and that △ABC is isosceles. ●

℗roof IV (indirect) In △ABC, assume m∠B > m∠C. $\overline{BE}$ and $\overline{DC}$ are the bisectors of ∠B and ∠C, respectively, and BE = DC.

Draw $\overline{BH} \parallel \overline{DC}$ and $\overline{CH} \parallel \overline{DB}$; then draw EH, as in Figure 5-5. Quadrilateral DCHB is a parallelogram. Therefore $\overline{BH} \cong \overline{DC} \cong \overline{BE}$, making △BHE isosceles. Thus:

$$m\angle BEH = m\angle BHE \qquad (I)$$

From our assumption that m∠B > m∠C:

$$m\angle CBE > m\angle BCD \quad \text{and} \quad CE > DB$$

Because CH = DB, we have CE > CH, which leads to:

$$m\angle CHE > m\angle CEH \qquad (II)$$

In △CEH, by adding (I) and (II), m∠BHC > m∠BEC. Because quadrilateral DCHB is a parallelogram, m∠BHC = m∠BDC.

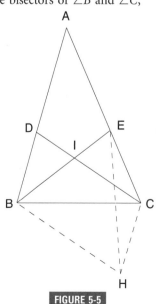

FIGURE 5-5

Thus, by substitution:

$$m\angle BDC > m\angle BEC$$

In $\triangle DBI$ and $\triangle ECI$:

$$m\angle DIB = m\angle EIC$$

Because $m\angle BDC > m\angle BEC$:

$$m\angle DBI < m\angle ECI$$

By doubling this inequality, we get $m\angle B < m\angle C$, thereby contradicting the assumption that $m\angle B > m\angle C$. A similar argument starting with the assumption that $m\angle B < m\angle C$ also leads to a contradiction. Thus we must conclude that $m\angle B = m\angle C$ and that $\triangle ABC$ is isosceles. ●

The following theorem is a direct consequence of Theorem 5.1.

| **THEOREM 5.2** | In a triangle, if two angles have unequal measures, the angle of greater measure has the shorter angle bisector. |

Proof   In Figure 5-6, $\triangle ABC$ has $m\angle ABC > m\angle ACB$. $\overline{BN}$ and $\overline{CK}$ are the angle bisectors of $\angle ABC$ and $\angle ACB$, respectively, and intersect at point I. Draw $\overline{BD}$ so that $m\angle DBN = m\angle ACK$. (Note: $\overline{BD}$ intersects $\overline{CK}$ at E.)

$\triangle DBN \sim \triangle DCE$ (AA), which yields $\dfrac{BD}{CD} = \dfrac{BN}{CE}$. Because $m\angle ABC > m\angle ACB$:

$$\frac{1}{2}\,m\angle ABC > \frac{1}{2}\,m\angle ACB \quad \text{or} \quad m\angle NBC > m\angle BCK$$

By construction, $m\angle DBN = m\angle ACK$. Therefore, by addition, $m\angle DBC > m\angle DCB$.
In $\triangle DBC$, $BD < CD$. From the above proportion we get $BN < CE$. Therefore $BN < CK$, our desired result. ●

INTERACTIVE 5-6

Drag *A*, *B*, and *C* to change the shape of the triangle and see that the larger angle always has the shorter bisector.

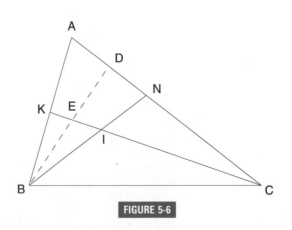

FIGURE 5-6

We have already established the various concurrency relationships involving angle bisectors (see Chapter 2). We now turn our attention to the measure of the angle formed by two interior angle bisectors of a triangle.

▌ THEOREM 5.3   The measure of the angle formed by two interior angle bisectors of a triangle equals the sum of the measure of a right angle and one-half the measure of the third angle of the triangle.

Ⓟroof   In Figure 5-7, the angle bisectors $\overline{BN}$ and $\overline{CM}$ intersect at point I.

INTERACTIVE 5-7

Drag *A, B,* and *C* to change the shape of the triangle and see that the formula is true.

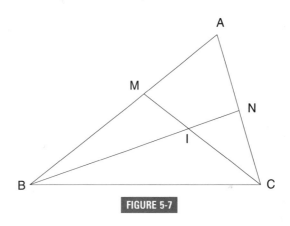

FIGURE 5-7

Consider $\triangle BIC$:

$$m\angle BIC = 180° - m\angle IBC - m\angle ICB$$

Then:

$$m\angle BIC = 180° - \frac{1}{2}(m\angle ABC) - \frac{1}{2}(m\angle ACB)$$

Because $m\angle ABC + m\angle ACB = 180° - m\angle A$, it follows that:

$$\frac{1}{2}(m\angle ABC) + \frac{1}{2}(m\angle ACB) = 90° - \frac{1}{2}(m\angle A)$$

By substitution:

$$m\angle BIC = 180° - [90 - \frac{1}{2}(m\angle A)] \quad \text{or} \quad m\angle BIC = 90° + \frac{1}{2}(m\angle A) \; ●$$

The natural extension of Theorem 5.3 involves *exterior* angle bisectors and is stated as Theorem 5.4.

Drag *A, B,* and *C* to change the shape of the triangle and see that the formula is true.

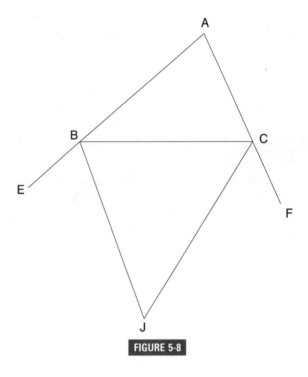

FIGURE 5-8

THEOREM 5.4 The measure of the angle formed by two exterior angle bisectors of a triangle equals the measure of a right angle minus one-half the measure of the third angle of the triangle.

Proof Figure 5-8 shows $\triangle ABC$ with exterior angle bisectors $\overline{BJ}$ and $\overline{CJ}$.

$$m\angle BJC = 180° - \frac{1}{2}(m\angle EBC) - \frac{1}{2}(m\angle FCB)$$

$$= 180° - \frac{1}{2}(180° - m\angle ABC) - \frac{1}{2}(180° - m\angle ACB)$$

$$= 180° - 90° + \frac{1}{2}(m\angle ABC) - 90° + \frac{1}{2}(m\angle ACB)$$

$$= \frac{1}{2}(m\angle ABC + m\angle ACB)$$

$$= \frac{1}{2}(180° - m\angle A)$$

$$m\angle BJC = 90° - \frac{1}{2}(m\angle A) \;\bullet$$

Our continuing study of angle bisectors now leads us to an investigation of the length of an angle bisector of a triangle. Specifically, we seek to find an expression relating the length of an angle bisector to the lengths of the sides (or their parts) of the triangle. This relationship is stated as Theorem 5.5.

THEOREM 5.5   In any triangle, the square of the length of the interior bisector of any angle is equal to the product of the lengths of the sides forming the bisected angle decreased by the product of the lengths of the segments of the side to which this bisector is drawn.

INTERACTIVE 5-9

Drag *A*, *B*, and *C* to change the shape of the triangle and see that the formula is true.

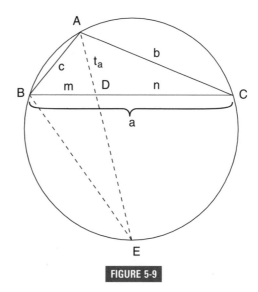

FIGURE 5-9

Proof   In Figure 5-9, $\overline{AD}$ (also labeled t_a) is the bisector of $\angle BAC$. Extend $\overline{AD}$ beyond point *D* to meet the circumcircle of $\triangle ABC$ at point *E*. Then draw $\overline{BE}$.

Because $m\angle BAD = m\angle CAD$ and $m\angle E = m\angle C$ (both angles are inscribed in the same arc):

$$\triangle ABE \sim \triangle ADC \Rightarrow \frac{AC}{AD} = \frac{AE}{AB}$$

Therefore:

$$(AC)(AB) = (AD)(AE) = (AD)(AD + DE) = (AD)^2 + (AD)(DE) \quad \text{(I)}$$

However:

$$(AD)(DE) = (BD)(DC) \quad\quad\quad\quad \text{(II)}$$

Substituting (II) into (I), we obtain:

$$(AD)^2 = (AC)(AB) - (BD)(DC)$$

Using the letter designations in Figure 5-9, we have:

$$t_a^2 = cb - mn \ \bullet$$

The following application illustrates the use of Theorem 5.5.

Application 1   The two shorter sides of a triangle measure 9 and 18. If the interior angle bisector drawn to the longest side measures 8, find the measure of the longest side of the triangle. ●

Solution   Let $AB = 9$, $AC = 18$, and the angle bisector $AD = 8$ (see Figure 5-10). Because $\dfrac{BD}{DC} = \dfrac{AB}{AC} = \dfrac{1}{2}$, we can let $BD = m = x$ so that $DC = n = 2x$.

From Theorem 5.5, we know:

$$t_a^2 = cb - mn \quad \text{or} \quad (AD)^2 = (AC)(AB) - (BD)(DC)$$

Therefore:

$$(8)^2 = (18)(9) - 2x^2 \quad \text{and} \quad x = 7$$

Thus:

$$BC = 3x = 21 \ ●$$

Suppose $\overline{AD}$ in Application 1 were *not* an angle bisector but rather just a nonspecific Cevian (i.e., a line segment joining a point on the side of a triangle with the opposite vertex). How would you then solve the problem? Would more information be necessary? Read on, and the answers to these questions will become apparent.

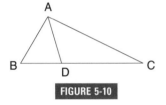

FIGURE 5-10

STEWART'S THEOREM

Essentially our problem is to find the length of "any" Cevian, a segment that has one endpoint on a vertex of a given triangle and the other endpoint on the opposite side. That is, if for $\triangle ABC$ (Figure 5-11) we know the lengths of $\overline{AC}$, $\overline{BC}$, $\overline{AD}$, and $\overline{BD}$, our problem is to find the length of $\overline{CD}$.

This problem was first solved by the famous Scottish geometer Robert Simson, who presented it in lectures but allowed his notes to be used by his prize student, Matthew Stewart, in his famous publication *General Theorems of Considerable Use in the Higher Parts of Mathematics* (Edinburgh, 1746). Simson's generosity was motivated by his desire to see Stewart obtain the chair of mathematics at the University of Edinburgh. He was successful. It is interesting to note how Simson was credited with a theorem he did not know (Theorem 3.7) yet was not credited with a theorem that he deserved to have credited to him (Theorem 5.6). We will refer to Theorem 5.6 by the author (Stewart) of the book in which it appeared.

INTERACTIVE 5-11

Drag *A*, *B*, and *C* to change the shape of the triangle and move point *D* and see that Stewart's theorem holds.

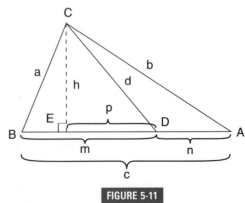

FIGURE 5-11

In fact, Simson deserves particular note for his definitive book *The Elements of Euclid* (Glasgow, 1756), which for over 150 years was published by other publishers as well. This book is the basis for all subsequent study of Euclid's *Elements*, including the high school geometry courses taught in the United States today.

We will first state Stewart's theorem, then prove it and provide some applications.

┃ THEOREM 5.6 **(Stewart's theorem)** Using the letter designation in Figure 5-11, the following relationship holds: $a^2n + b^2m = c(d^2 + mn)$.

---

❶roof In $\triangle ABC$, let $BC = a$, $AC = b$, $AB = c$, and $CD = d$. Point D divides $\overline{AB}$ into two segments: $BD = m$ and $DA = n$. Draw altitude $CE = h$ and let $ED = p$.

In order to proceed with the proof of Stewart's theorem, we must first derive two formulas. The first is applicable to $\triangle CBD$. We apply the Pythagorean theorem to $\triangle CEB$ to obtain $(CB)^2 = (CE)^2 + (BE)^2$. Because $BE = m - p$:

$$a^2 = h^2 + (m - p)^2 \qquad\text{(I)}$$

By applying the Pythagorean theorem to $\triangle CED$, we have:

$$(CD)^2 = (CE)^2 + (ED)^2 \quad\text{or}\quad h^2 = d^2 - p^2$$

Substituting for h^2 in equation (I), we obtain:

$$\begin{aligned}
a^2 &= d^2 - p^2 + (m - p)^2 \\
&= d^2 - p^2 + m^2 - 2mp + p^2 \\
a^2 &= d^2 + m^2 - 2mp \qquad\text{(II)}
\end{aligned}$$

A similar argument is applicable to $\triangle CDA$. Applying the Pythagorean theorem to $\triangle CEA$, we find that $(CA)^2 = (CE)^2 + (EA)^2$. Because $EA = (n + p)$:

$$b^2 = h^2 + (n + p)^2 \qquad\text{(III)}$$

However, $h^2 = d^2 - p^2$, so we substitute for h^2 in equation (III) as follows:

$$\begin{aligned}
b^2 &= d^2 - p^2 + (n + p)^2 \\
&= d^2 - p^2 + n^2 + 2np + p^2 \\
b^2 &= d^2 + n^2 + 2np \qquad\text{(IV)}
\end{aligned}$$

Equations (II) and (IV) give us the formulas we need. Multiply equation (II) by n to get:

$$a^2n = d^2n + m^2n - 2mnp \qquad\text{(V)}$$

Now multiply equation (IV) by m to get:

$$b^2m = d^2m + n^2m + 2mnp \qquad\text{(VI)}$$

Adding (V) and (VI), we have:

$$a^2n + b^2m = d^2n + d^2m + m^2n + n^2m + 2mnp - 2mnp$$

Therefore $a^2n + b^2m = d^2(n + m) + mn(m + n)$. Because $m + n = c$, we have:

$$a^2n + b^2m = d^2c + mnc \quad \text{or} \quad a^2n + b^2m = c(d^2 + mn)$$

This is the relationship we set out to develop. ●

Stewart's theorem can be applied to a variety of situations, some of which we offer here.

Application 2   In an isosceles triangle with two congruent sides that measure 17, a line measuring 16 is drawn from the vertex to the base. If one segment of the base, as cut by this line, exceeds the other by 8, find the lengths of the two segments. ●

Solution   In Figure 5-12, $AB = AC = 17$ and $AD = 16$. Let $BD = x$; therefore $DC = x + 8$. By Stewart's theorem,

$$(AB)^2(DC) + (AC)^2(BD) = BC[(AD)^2 + (BD)(DC)]$$

Therefore:

$$(17)^2(x + 8) + (17)^2(x) = (2x + 8)[(16)^2 + x(x + 8)] \quad \text{and} \quad x = 3$$

Therefore $BD = 3$ and $DC = 11$. ●

Application 3   Prove that in a right triangle the sum of the squares of the distances from the vertex of the right angle to the trisection points along the hypotenuse is equal to five-ninths the square of the measure of the hypotenuse. ●

Proof   Applying Stewart's theorem to Figure 5-13, using p as the internal line segment, we find that:

$$2a^2n + b^2n = c(p^2 + 2n^2) \tag{I}$$

Using q as the internal line segment:

$$a^2n + 2b^2n = c(q^2 + 2n^2) \tag{II}$$

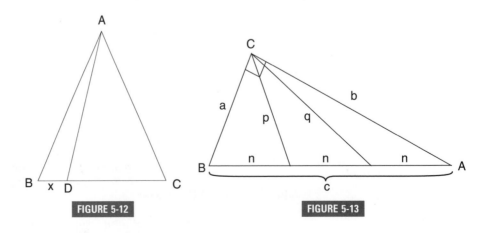

FIGURE 5-12   FIGURE 5-13

Adding (I) and (II), we get:

$$3a^2n + 3b^2n = c(4n^2 + p^2 + q^2)$$

Because $a^2 + b^2 = c^2$:

$$3n(c^2) = c(4n^2 + p^2 + q^2)$$

Because $3n = c$:

$$c^2 = (2n)^2 + p^2 + q^2$$

But $2n = \dfrac{2}{3}c$; therefore:

$$p^2 + q^2 = c^2 - \left(\frac{2}{3}c\right)^2 = \frac{5}{9}c^2 \ \bullet$$

Ⓐpplication 4   To illustrate the power of Stewart's theorem, we offer another proof of Theorem 5.1.*
This straightforward method takes this "elementary" theorem and places it (temporarily) at a more advanced point in the development of Euclidean geometry. ●

Ⓟroof   Let $\overline{BE}$ and $\overline{CD}$ be angle bisectors in $\triangle ABC$, with $BE = CD = x$ (Figure 5-14).
We need to show that $b = c$. An angle bisector divides the side it is drawn to
into segments of lengths proportional to the two other sides of the triangle.
Thus:

$$BD = \frac{ac}{a + b} \qquad AD = \frac{bc}{a + b} \qquad AE = \frac{bc}{a + c} \qquad CE = \frac{ab}{a + c}$$

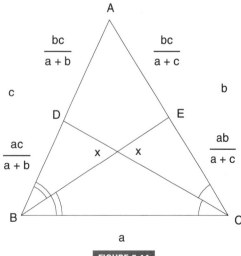

FIGURE 5-14

---

* This proof was contributed by Jan Siwanowicz.

Applying Stewart's theorem twice to $\triangle ABC$, we obtain:

$$a^2 \frac{bc}{a+c} + c^2 \frac{ab}{a+c} = b\left(x^2 + \frac{bc}{a+c} \cdot \frac{ab}{a+c}\right)$$

$$a^2 \frac{bc}{a+b} + b^2 \frac{ac}{a+b} = c\left(x^2 + \frac{bc}{a+b} \cdot \frac{ac}{a+b}\right)$$

Solving for x^2, we obtain:

$$x^2 = ac - \frac{ab^2c}{(a+c)^2} = ab - \frac{abc^2}{(a+b)^2}$$

Thus:

$$c + \frac{bc^2}{(a+b)^2} = b + \frac{b^2c}{(a+c)^2}$$

This may be expressed simply as:

$$c\left(1 + \frac{bc}{(a+b)^2}\right) = b\left(1 + \frac{bc}{(a+c)^2}\right)$$

If $b > c$, because a, b, $c > 0$ we have:

$$\left(1 + \frac{bc}{(a+b)^2}\right) < \left(1 + \frac{bc}{(a+c)^2}\right)$$

Thus equality (I) does not hold.

If $b < c$, we have:

$$\left(1 + \frac{bc}{(a+b)^2}\right) > \left(1 + \frac{bc}{(a+c)^2}\right)$$

Again equality (I) does not hold.

Thus $b = c$, which completes the proof. ●

MIQUEL'S THEOREM

You might want to try this experiment. Draw any convenient triangle and select a point on each side. Now construct three circles, each containing two of these points and the vertex determined by the two sides on which these points lie. Although you can do this on paper with the aid of a pair of compasses, it is particularly nice to do this with The Geometer's Sketchpad. What relationship do you notice about these three circles? Your observation should lead you to a theorem published by A. Miquel in 1838. We will state this theorem as follows.

THEOREM 5.7 **(Miquel's theorem)** If a point is selected on each side of a triangle, then the circles determined by each vertex and the points on the adjacent sides pass through a common point.

This theorem can be viewed in two ways. The expected form is shown in Figure 5-15. However, when two of the selected points are on the extensions of the sides, the theorem still holds. This form is shown in Figure 5-16.

INTERACTIVE 5-15

Drag *A*, *B*, and *C* to change the shape of the triangle and *D*, *E*, and *F* on the sides and see that the circles share a common point.

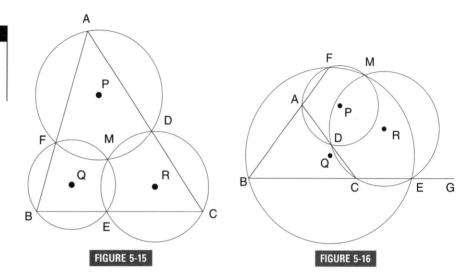

FIGURE 5-15   **FIGURE 5-16**

Proof **Case I** Consider the problem when point *M* is inside $\triangle ABC$, as shown in Figure 5-17. Points *D*, *E*, and *F* are any points on sides $\overline{AC}$, $\overline{BC}$, and $\overline{AB}$, respectively, of $\triangle ABC$. Let circles *Q* and *R*, determined by points *F*, *B*, *E* and *D*, *C*, *E*, respectively, meet at point *M*. Draw $\overline{FM}$, $\overline{ME}$, and $\overline{MD}$.

In cyclic quadrilateral *BFME*:

$$m\angle FME = 180° - m\angle B$$

Similarly, in cyclic quadrilateral *CDME*:

$$m\angle DME = 180° - m\angle C$$

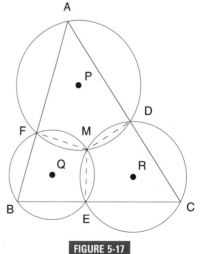

FIGURE 5-17

By addition:

$$m\angle FME + m\angle DME = 360° - (m\angle B + m\angle C)$$

Therefore:

$$m\angle FMD = m\angle B + m\angle C$$

However, in $\triangle ABC$:

$$m\angle B + m\angle C = 180° - m\angle A$$

Therefore $m\angle FMD = 180° - m\angle A$ and quadrilateral $AFMD$ is cyclic. Thus point M lies on all three circles. ●

Case II Figure 5-18 illustrates the problem when point M is outside $\triangle ABC$. Again let circles Q and R meet at point M. Because quadrilateral $BFME$ is cyclic:

$$m\angle FME = 180° - m\angle B$$

Similarly, because quadrilateral $CDME$ is cyclic:

$$m\angle DME = 180° - m\angle DCE$$

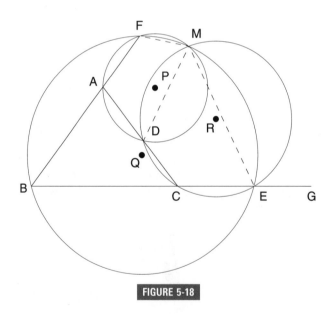

FIGURE 5-18

By subtraction:

$$m\angle FMD = m\angle FME - m\angle DME = m\angle DCE - m\angle B \qquad \text{(I)}$$

However:

$$m\angle DCE = m\angle BAC + m\angle B \qquad \text{(II)}$$

By substituting (II) into (I), we get:

$$m\angle FMD = m\angle BAC = 180° - m\angle FAD$$

Therefore quadrilateral *ADMF* is also cyclic, and point *M* lies on all three circles. ●

Point *M* is called the *Miquel point* of △*ABC*. The points *F*, *D*, and *E* determine the *Miquel triangle*, △*FDE*. Miquel's theorem opens the door to a variety of additional theorems. We present some of them here.

▌ THEOREM 5.8   The segments joining the Miquel point of a triangle to the vertices of the Miquel triangle form congruent angles with the respective sides of the original triangle.

℗roof   Because quadrilateral *AFMD* is cyclic (see Figures 5-17 and 5-18), ∠*AFM* is supplementary to ∠*ADM*. But ∠*ADM* is supplementary to ∠*CDM*. Therefore ∠*AFM* ≅ ∠*CDM*, whereupon it follows that ∠*BFM* ≅ ∠*ADM*. To complete the proof, merely apply the same argument to cyclic quadrilateral *CDME*. ●

We say that a triangle is inscribed in a second triangle if each of the vertices of the first triangle lies on the sides of the second triangle. Thus we state the following theorem.

▌ THEOREM 5.9   Two triangles inscribed in the same triangle and having a common Miquel point are similar.

℗roof   Consider △*DEF* and △*D′E′F′*, which have the same Miquel point *M* (see Figure 5-19). From Theorem 5.8, we find that:

$$\angle MFB \cong \angle MDA$$
$$\angle MF'A \cong \angle MD'C$$

Therefore △*MF′F* ~ △*MD′D*. Similarly, △*MD′D* ~ △*ME′E*. Thus:

$$\angle FMF' \cong \angle DMD' \cong \angle EME'$$

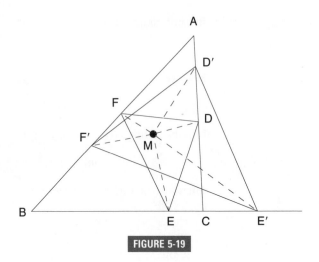

FIGURE 5-19

By addition:

$$\angle F'MD' \cong \angle FMD$$
$$\angle F'ME' \cong \angle FME$$
$$\angle E'MD' \cong \angle EMD$$

Also, as a result of the above similar triangles:

$$\frac{MF}{MF'} = \frac{MD}{MD'} = \frac{ME}{ME'}$$

Because two triangles are similar if two pairs of corresponding sides are proportional and the included angles are congruent, we get:

$$\triangle F'MD \sim \triangle FMD$$
$$\triangle F'ME' \sim \triangle FME$$
$$\triangle E'MD' \sim \triangle EMD$$

Therefore:

$$\frac{F'D'}{FD} = \frac{F'M}{FM} \quad \text{and} \quad \frac{F'E'}{FE} = \frac{F'M}{FM} \Rightarrow \frac{F'D'}{FD} = \frac{F'E'}{FE}$$

Similarly:

$$\frac{E'D'}{ED} = \frac{F'E'}{FE}$$

This proves that $\triangle DEF \sim \triangle D'E'F'$ because the corresponding sides are proportional. ●

THEOREM 5.10 The centers of Miquel circles of a given triangle determine a triangle similar to the given triangle.

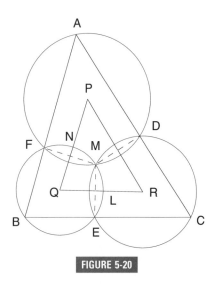

FIGURE 5-20

Proof Draw common chords $\overline{FM}$, $\overline{EM}$, and $\overline{DM}$. $\overline{PQ}$ meets circle Q at point N, and $\overline{RQ}$ meets circle Q at point L (see Figure 5-20).

Because the line of centers of two circles is the perpendicular bisector of their common chord, $\overline{PQ}$ is the perpendicular bisector of $\overline{FM}$, so $m\widehat{FN} = m\widehat{NM}$. Similarly, $\overline{QR}$ bisects $\widehat{EM}$, so $m\widehat{ML} = m\widehat{LE}$.

Now:

$$m\angle NQL = (m\widehat{NM} + m\widehat{ML}) = \frac{1}{2}(m\widehat{FE}) \quad \text{and} \quad m\angle FBE = \frac{1}{2}(m\widehat{FE})$$

Therefore:

$$m\angle NQL = m\angle FBE$$

In a similar fashion, it may be proved that $m\angle QPR = m\angle BAC$. Thus $\triangle PQR \sim \triangle ABC$. ●

Before completing this introductory study of Miquel's theorem, you will find it interesting to apply the theorem to an equilateral triangle as well as to special right triangles. Are there any new conclusions to be drawn?

MEDIANS

Although Stewart's theorem can certainly be applied to the medians of a triangle, there are many interesting properties of medians that are not direct results of Stewart's theorem. Some of these are certainly worthy of our attention.

When asked for a property of the medians of a triangle, the typical high school geometry student will probably be quick to respond that the point of intersection of the medians (the centroid, or center of gravity) is a trisection point of each median. In Chapter 2, we used Ceva's theorem to prove that the medians of a triangle are concurrent. The student may also recall that the median of a triangle partitions the triangle into two triangles of equal area. This property can easily be extended to a realization that the three medians of a triangle partition the triangle into six triangles of equal area.

Our first task will be to examine the relative lengths of the medians of a triangle. Using The Geometer's Sketchpad, draw a scalene triangle and its medians. Can you guess which median of the triangle is longest and which is shortest? Measure the medians using The Geometer's Sketchpad. Were your guesses right? Knowing the lengths of the sides of this given triangle, could you order the lengths of the medians *without measuring* them? This is what our next theorem does for us.

THEOREM 5.11	In a triangle, the longest side corresponds to the shortest median.

Proof

INTERACTIVE 5-21

Drag *A*, *B*, and *C* to change the shape of the triangle and see that the longest side corresponds to the shortest median.

Assume $AC > AB$. We must show that $BE < CD$ (see Figure 5-21). $\triangle AFB$ and $\triangle AFC$ are two triangles that have two sides of the same length (i.e., $BF = CF$ and $AF = AF$). Therefore, because $AC > AB$, $m\angle AFC > m\angle AFB$. $\triangle GFB$ and $\triangle GFC$ are also two triangles that have two sides of the same length. Therefore, because $m\angle GFC > m\angle GFB$, $GC > GB$. Because of the trisection property of the centroid, $CD > BE$. ●

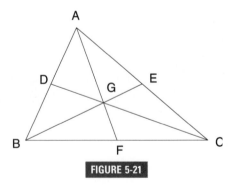

FIGURE 5-21

We can find the length of a median with Stewart's theorem, and we know (from Theorem 5.11) the relationship between the lengths of the medians with regard to the lengths of the sides of the triangle. We turn our attention in the next two theorems to some interesting relationships about the sum of the lengths of the medians of a given triangle.

THEOREM 5.12   For any triangle, the sum of the lengths of the medians is less than the perimeter of the triangle.

Proof $\overline{AF}$, $\overline{BE}$, and $\overline{CD}$ are the medians of $\triangle ABC$. Begin the proof by choosing point N on $\overrightarrow{AF}$ so that $AF = NF$ (see Figure 5-22). Quadrilateral $ACNB$ is a parallelogram because the diagonals bisect each other. Therefore $BN = AC$. For $\triangle ABN$, $AN < AB + BN$, which with appropriate substitutions gives us:

INTERACTIVE 5-22

Drag *A, B,* and *C* to change the shape of the triangle and see that the inequality is true.

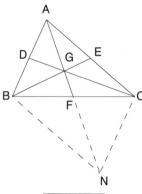

FIGURE 5-22

$$2(AF) < AB + AC \quad \text{or} \quad 2(m_a) < c + b$$

Similarly, we can show that:

$$2(m_b) < a + c \quad \text{and} \quad 2(m_c) < a + b$$

By addition:

$$2(m_a + m_b + m_c) < 2(a + b + c) \quad \text{or} \quad m_a + m_b + m_c < a + b + c \; \bullet$$

THEOREM 5.13   For any triangle, the sum of the lengths of the medians is greater than three-fourths the perimeter of the triangle.

Proof   We begin by using the trisection property of the centroid G of $\triangle ABC$ (see Figure 5-21). In $\triangle BGC$:

$$BG + CG > BC \quad \text{or} \quad \frac{2}{3}(m_c) + \frac{2}{3}(m_b) > a$$

In a similar way, we get:

$$\frac{2}{3}(m_a) + \frac{2}{3}(m_c) > b \quad \text{and} \quad \frac{2}{3}(m_a) + \frac{2}{3}(m_b) > c$$

By addition:

$$\frac{4}{3}(m_a + m_b + m_c) > a + b + c$$

Therefore:

$$m_a + m_b + m_c > \frac{3}{4}(a + b + c) \; \bullet$$

The previous two theorems tell us that $\frac{3}{4}(a + b + c) < m_a + m_b + m_c < a + b + c$.

We now turn our attention to the squares of the lengths of the medians of a given triangle.

THEOREM 5.14   Twice the square of the length of a median of a triangle equals the sum of the squares of the lengths of the two including sides minus one-half the square of the length of the third side.

Proof   By applying Stewart's theorem to $\triangle ABC$ in Figure 5-23, we get:

$$(AB)^2(FC) + (AC)^2(BF) = (BF + FC)[(AF)^2 + (BF)(FC)]$$

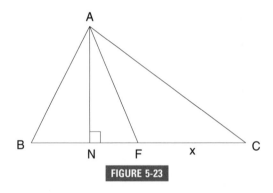

FIGURE 5-23

Let $x = FC = BF$. Then:

$$x(AB)^2 + x(AC)^2 = 2x[(AF)^2 + x^2]$$
$$(AB)^2 + (AC)^2 = 2[(AF)^2 + x^2]$$
$$2(AF)^2 = (AB)^2 + (AC)^2 - 2x^2$$

Because $x = \frac{1}{2}(BC)$, we obtain our desired result:

$$2(AF)^2 = (AB)^2 + (AC)^2 - \frac{1}{2}(BC)^2 \ \bullet$$

This theorem by itself is not too exciting, yet it helps us prove some rather useful and interesting properties, one of which we state as Theorem 5.15.

THEOREM 5.15 The sum of the squares of the lengths of the medians of a triangle equals three-fourths the sum of the squares of the lengths of the sides of the triangle.

Proof The proof of this theorem primarily uses the result stated in Theorem 5.14, namely:

$$2m_a^2 = b^2 + c^2 - \frac{1}{2}a^2$$

$$2m_b^2 = a^2 + c^2 - \frac{1}{2}b^2$$

$$2m_c^2 = a^2 + b^2 - \frac{1}{2}c^2$$

By addition:

$$2(m_a^2 + m_b^2 + m_c^2) = 2(a^2 + b^2 + c^2) - \frac{1}{2}(a^2 + b^2 + c^2)$$

$$2(m_a^2 + m_b^2 + m_c^2) = \frac{3}{2}(a^2 + b^2 + c^2)$$

$$m_a^2 + m_b^2 + m_c^2 = \frac{3}{4}(a^2 + b^2 + c^2)$$

This is our desired result. ●

We may immediately use this result to establish a relationship between the sum of the squares of the lengths of the segments joining the centroid with the vertices and the sum of the squares of the lengths of the sides.

THEOREM 5.16 The sum of the squares of the lengths of the segments joining the centroid with the vertices is one-third the sum of the squares of the lengths of the sides.

Proof The length of a segment joining the centroid with a vertex is two-thirds the length of its respective median. We therefore seek to find:

$$\left(\frac{2}{3}m_a\right)^2 + \left(\frac{2}{3}m_b\right)^2 + \left(\frac{2}{3}m_c\right)^2 = \frac{4}{9}(m_a^2 + m_b^2 + m_c^2)$$

From Theorem 5.15, we have:

$$m_a^2 + m_b^2 + m_c^2 = \frac{3}{4}(a^2 + b^2 + c^2)$$

Therefore:

$$\frac{4}{9}(m_a^2 + m_b^2 + m_c^2) = \frac{1}{3}(a^2 + b^2 + c^2)$$

This relationship is what was to be demonstrated. ●

The next theorem is more general—it relates *any* point in the plane of a triangle to segments of the triangle.

> ▌**THEOREM 5.17** If P is any point in the plane of $\triangle ABC$ with centroid G, then $(AP)^2 + (BP)^2 + (CP)^2 = (AG)^2 + (BG)^2 + (CG)^2 + 3(PG)^2$ (see Figure 5-24).

INTERACTIVE 5-24

Drag *A*, *B*, and *C* to change the shape of the triangle and change the position of *P* and see that the equation is true.

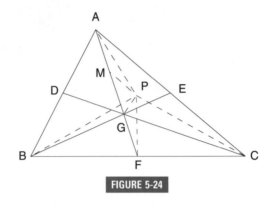

FIGURE 5-24

Proof   Begin by letting M be the midpoint of $\overline{AG}$ (see Figure 5-24). We now apply Theorem 5.14 to each triangle indicated:

$$\triangle PBC\colon 2(PF)^2 = (BP)^2 + (CP)^2 - \frac{1}{2}(BC)^2 \tag{I}$$

$$\triangle PAG\colon 2(PM)^2 = (AP)^2 + (PG)^2 - \frac{1}{2}(AG)^2 \tag{II}$$

$$\triangle PMF\colon 2(PG)^2 = (PM)^2 + (PF)^2 - \frac{1}{2}(MF)^2 \tag{III}$$

Because $MF = \frac{2}{3}(AF)$ and $AG = \frac{2}{3}(AF)$, $MF = AG$.

Substituting into equation (III) and multiplying by 2, we get:

$$4(PG)^2 = 2(PM)^2 + 2(PF)^2 - (AG)^2 \tag{IV}$$

Adding (I), (II), and (IV):

$$2(PF)^2 + 2(PM)^2 + 4(PG)^2 = (BP)^2 + (AP)^2 + 2(PM)^2 + (CP)^2 + (PG)^2$$
$$+ 2(PF)^2 - \frac{1}{2}(BC)^2 - \frac{1}{2}(AG)^2 - (AG)^2$$

$$(AP)^2 + (BP)^2 + (CP)^2 - 3(PG)^2 = \frac{3}{2}(AG)^2 + \frac{1}{2}(BC)^2 \tag{V}$$

A similar argument made for median $\overline{BE}$ yields:

$$(AP)^2 + (BP)^2 + (CP)^2 - 3(PG)^2 = \frac{3}{2}(BG)^2 + \frac{1}{2}(AC)^2 \qquad (VI)$$

For median $\overline{CD}$, we get:

$$(AP)^2 + (BP)^2 + (CP)^2 - 3(PG)^2 = \frac{3}{2}(CG)^2 + \frac{1}{2}(AB)^2 \qquad (VII)$$

Adding (V), (VI), and (VII):

$$3[(AP)^2 + (BP)^2 + (CP)^2 - 3(PG)^2] = \frac{3}{2}[(AG)^2 + (BG)^2 + (CG)^2]$$
$$+ \frac{1}{2}[(BC)^2 + (AC)^2 + (AB)^2] \qquad (VIII)$$

We now apply Theorem 5.16 to $\triangle ABC$:

$$(AG)^2 + (BG)^2 + (CG)^2 = \frac{1}{3}[(BC)^2 + (AC)^2 + (AB)^2] \quad \text{or}$$
$$3[(AG)^2 + (BG)^2 + (CG)^2] = (BC)^2 + (AC)^2 + (AB)^2$$

We substitute this into equation (VIII) to get our desired result:

$$3[(AP)^2 + (BP)^2 + (CP)^2 - 3(PG)^2] = \frac{3}{2}[(AG)^2 + (BG)^2 + (CG)^2]$$
$$+ \frac{1}{2}(3[(AG)^2 + (BG)^2 + (CG)^2])$$

$$(AP)^2 + (BP)^2 + (CP)^2 = (AG)^2 + (BG)^2 + (CG)^2 + 3(PG)^2 \bullet$$

The medians of a triangle provide us with many interesting relationships. We will investigate some now and leave others as exercises.

THEOREM 5.18 In any triangle, a median and the midline that intersects it (in the interior of the triangle) bisect each other.

Proof We wish to prove that median $\overline{AF}$ and midline $\overline{DE}$ bisect each other (see Figure 5-25). By drawing midlines $\overline{DF}$ and $\overline{EF}$, we form parallelogram $ADFE$ (opposite sides are parallel). Therefore diagonals $\overline{AF}$ and $\overline{DE}$ bisect each other. $\bullet$

The centroid serves as a sort of "balancing" point of a triangle. We examine this property in our next theorem.

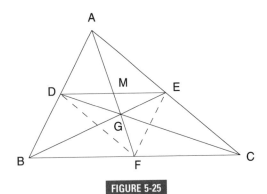

FIGURE 5-25

INTERACTIVE 5-26

Drag *A, B,* and *C* to change the
shape of the triangle and
change the position of *G* and see
that the equation is true.

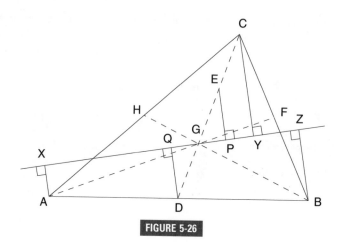

FIGURE 5-26

THEOREM 5.19   In any triangle *ABD*, let $\overleftrightarrow{XYZ}$ be any line through the centroid *G*. If perpendiculars are drawn from each vertex of $\triangle ABC$ to this line, as shown in Figure 5-26, then $CY = AX + BZ$.

Proof   Draw medians $\overline{CD}$, $\overline{AF}$, and $\overline{BH}$ (see Figure 5-26). From *E*, the midpoint of $\overline{CG}$, draw $\overline{EP} \perp \overline{XZ}$. Also draw $\overline{DQ} \perp \overline{XZ}$. Because $\angle CGY \cong \angle QGD$ and $EC = EG = DG$ (property of a centroid):

$$\triangle QGD \cong \triangle PGE \quad \text{and} \quad QD = EP$$

$\overline{AX} \parallel \overline{BZ}$; therefore $\overline{QD}$ is the median of trapezoid *AXZB* and:

$$QD = \frac{1}{2}(AX + BZ) \quad \text{(property of median of a trapezoid)}$$

Also:

$$EP = \frac{1}{2}(CY) \quad \text{(property of a midline)}$$

Therefore, by transitivity:

$$\frac{1}{2}(CY) = \frac{1}{2}(AX + BZ) \quad \text{or} \quad CY = AX + BZ \;\bullet$$

It is interesting to note that for a given point in a given circle an infinite number of inscribed triangles exist that have this point as a centroid. We state this property as Theorem 5.20.

THEOREM 5.20   An infinite number of triangles each having a given interior point as centroid can be inscribed in a given circle.

Proof

INTERACTIVE 5-27

Choose any interior point as a centroid and note that a triangle can be inscribed in the given circle.

This proof will be somewhat different from others we have used so far. To show that there exist an infinite number of triangles with the necessary specifications, we will show that one such triangle, *randomly selected*, exists. This will imply that an infinite number of other triangles similarly constructed also exist.

We begin by selecting any point on circle O. This point will be point A of $\triangle ABC$ (see Figure 5-27). Join point A with given centroid G and extend $\overline{AG}$ through G to point F, so that $GF = \frac{1}{2}(AG)$. Then draw $\overline{OF}$. At point F, construct a perpendicular to $\overline{OF}$, intersecting the circle at points B and C.

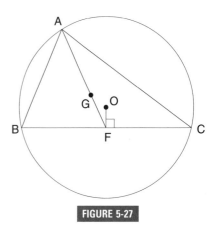

FIGURE 5-27

This easily justifiable construction proves that a triangle exists with the specified conditions. But because many other triangles can similarly be constructed (depending on the selection of point A), our proof is complete. ●

We conclude our study of the medians of a triangle by briefly considering the *medial triangle,* that is, the triangle formed by joining the midpoints of the sides of a triangle.

THEOREM 5.21   A triangle and its medial triangle have the same centroid.

Proof   In $\triangle ABC$, median $\overline{AF}$ bisects $\overline{DE}$ at point M (Theorem 5.18). Therefore $\overline{FM}$ is a median of medial triangle DEF (see Figure 5-28). Similarly, $\overline{DK}$ and $\overline{EN}$ are medians of $\triangle DEF$ as well as being segments of medians of $\triangle ABC$. Because the medians of $\triangle ABC$ meet at point G, so do the medians of $\triangle DEF$. ●

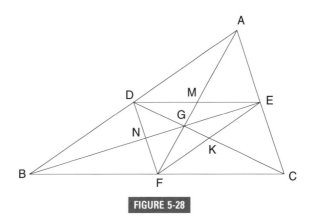

FIGURE 5-28

In looking back over this chapter, you will see that we began with a study of angle bisectors of a triangle. We then considered a general segment of a triangle (a Cevian) to exhibit the usefulness of Stewart's theorem. Finally, we studied properties of the medians of a triangle. Your knowledge of triangles should now be considerably more extensive.

EXERCISES

1. Prove that the sum of the reciprocals of the lengths of the interior angle bisectors of a triangle is greater than the sum of the reciprocals of the lengths of the sides of the triangle.

2. Prove that the feet of the four perpendiculars drawn from a vertex of a triangle to the two interior and two exterior angle bisectors of the other two angles of the triangle are collinear.

3. Prove that the difference of the measures of the two angles that an interior angle bisector forms with the opposite side equals the difference of the measures of the two remaining angles of the triangle.

4. Prove that the measure of the angle formed by the exterior angle bisector and the opposite side of the triangle equals one-half the difference of the measures of the two remaining angles of the triangle.

5. In a 30-60-90 triangle with hypotenuse of length 4, find the distance from the vertex of the right angle to the point of intersection of the angle bisectors.

6. In a right triangle, the bisector of the right angle divides the hypotenuse into segments that measure 3 and 4. Find the measure of the angle bisector of the larger acute angle of the right triangle.

7. Use Stewart's theorem to find the length of the medians of a triangle in terms of the lengths of its sides and their segments.

8. Prove that any triangle whose sides contain the vertices of a Miquel triangle of a given triangle and whose vertices each lie on a Miquel circle is similar to the given triangle.

9. Prove that two similar triangles inscribed in the same triangle have the same Miquel point.

10. Using Figure 5-17, prove that $m\angle BMC = m\angle BAC + m\angle FED$.

11. Prove that if three circles have a common point of intersection, M, then there are three or more similar triangles for which M is the Miquel point.

12. Prove that if a triangle is constructed with sides the length of the medians of a given triangle, the lengths of the medians of this newly constructed triangle are each three-fourths the length of the respective sides of the given triangle.

13. Prove that the area of a triangle whose sides are the length of the medians of a given triangle is equal to three-fourths the area of the given triangle.

14. Prove that if two points are equidistant from the centroid of a triangle, then the sums of the squares of their distances from the vertices of the triangle are equal.

15. Prove that the line containing the midpoint of a median of a triangle and a vertex (not on the median) trisects a side of the triangle.

16. Prove that the medians of a triangle partition the triangle into six triangles of equal area.

17. Prove that the lines containing the vertices of a triangle and parallel to the opposite sides form a new triangle that has the original triangle as its medial triangle.

18. Prove that for a right triangle $\dfrac{1}{h_c^2} = \dfrac{1}{h_a^2} + \dfrac{1}{h_b^2}$. Then prove the converse.

19. Prove that for a right triangle $5m_c^2 = m_a^2 + m_b^2$. Then prove the converse.

CHAPTER SIX

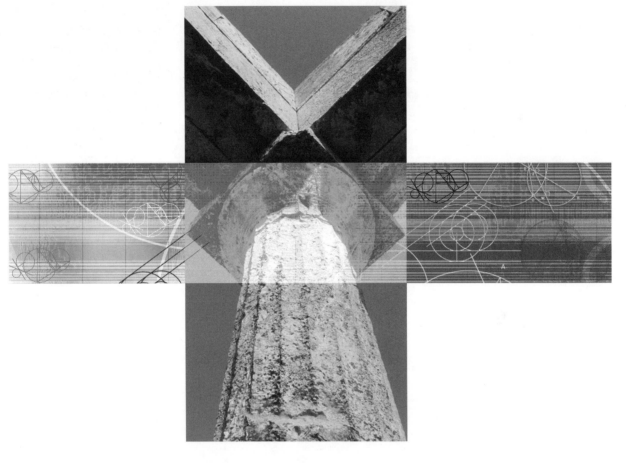

QUADRILATERALS

We begin our study of quadrilaterals where high school geometry leaves off. Most of the study on quadrilaterals in elementary geometry deals with special quadrilaterals such as trapezoids, parallelograms, rhombuses, rectangles, and squares. Let us look first at the general quadrilateral, that is, one with no special properties, and then at the cyclic quadrilateral, that is, one that can be inscribed in a circle.

Suppose you were to draw any shape quadrilateral and then join (with segments) the midpoints of consecutive sides. What would you expect the resulting quadrilateral to look like? The Geometer's Sketchpad is very helpful in our experiment. Construct a quadrilateral, locate and join the midpoints of the sides, and then distort the original quadrilateral, observing the shape of the quadrilateral formed by joining the midpoints of the sides of the original quadrilateral. What you will readily notice is stated as our first theorem of this chapter.

| **THEOREM 6.1** | The quadrilateral formed by joining the midpoints of consecutive sides of any quadrilateral is a parallelogram. |

Proof   In Figure 6-1, points P, Q, R, and S are the midpoints of the sides of quadrilateral $ABCD$. In $\triangle ADB$, $\overline{PQ}$ is a midline, and therefore:

INTERACTIVE 6-1

Drag points A, B, C, and D to change the shape of the quadrilateral and see that $PQRS$ is always a parallelogram.

$$\overline{PQ} \parallel \overline{DB} \quad \text{and} \quad PQ = \frac{1}{2}(DB)$$

In $\triangle CDB$, $\overline{SR}$ is a midline, and therefore:

$$\overline{SR} \parallel \overline{DB} \quad \text{and} \quad SR = \frac{1}{2}(DB)$$

Thus $\overline{PQ} \parallel \overline{SR}$ and $PQ = SR$, which establishes that quadrilateral $PQRS$ is a parallelogram. ●

The question that now arises is, What type of quadrilateral $ABCD$ will produce a rectangle $PQRS$, a rhombus $PQRS$, or a square $PQRS$?

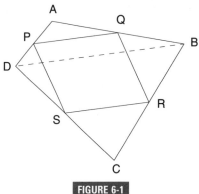

FIGURE 6-1

THEOREM 6.2 The quadrilateral formed by joining the midpoints of consecutive sides of a quadrilateral whose diagonals are perpendicular is a rectangle.

Because $\overline{QR} \parallel \overline{AC}$ in Figure 6-1, quadrilateral $PQRS$ would be a rectangle (i.e., a parallelogram with adjacent sides perpendicular) if $\overline{PQ} \perp \overline{QR}$. This is true if $\overline{AC} \perp \overline{DB}$.

THEOREM 6.3 The quadrilateral formed by joining the midpoints of consecutive sides of a quadrilateral whose diagonals are congruent is a rhombus.

INTERACTIVE 6-2

Drag points *A, B, C,* and *D* to change the shape of the quadrilateral and see that *PQRS* is always a rhombus.

Proof Suppose we have a quadrilateral with congruent diagonals (see Figure 6-2). The midline $\overline{PQ}$ of $\triangle ABD$ has the property:

$$PQ = \frac{1}{2}(BD)$$

Similarly, for $\triangle ABC$ and midline $\overline{QR}$:

$$QR = \frac{1}{2}(AC)$$

Because in this quadrilateral $BD = AC$, we have:

$$PQ = QR$$

Thus parallelogram $PQRS$ is a rhombus. ●

Combining the results of Theorems 6.2 and 6.3 enables us to establish the next theorem.

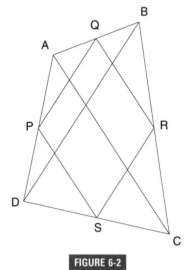

FIGURE 6-2

THEOREM 6.4 The quadrilateral formed by joining the midpoints of consecutive sides of a quadrilateral whose diagonals are perpendicular and congruent is a square.

CENTERS OF A QUADRILATERAL

We will now consider two centers of a quadrilateral. The *centroid* of a quadrilateral is that point on which a quadrilateral of uniform density will balance. This point may be found in the following way. Let points M and N be the centroids of $\triangle ABC$ and $\triangle ADC$, respectively (see Figure 6-3). Let points K and L be the centroids of $\triangle ABD$ and $\triangle BCD$, respectively. The point of intersection, G, of $\overline{MN}$ and $\overline{KL}$ is the centroid of quadrilateral *ABCD*.

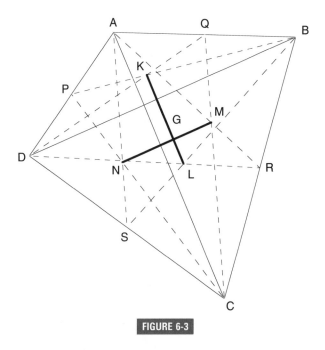

FIGURE 6-3

The *centerpoint* of a quadrilateral is the point of intersection of the two segments joining the midpoints of the opposite sides of the quadrilateral. In Figure 6-4, point G is the *centerpoint* of quadrilateral *ABCD*.

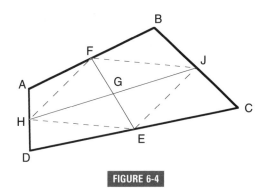

FIGURE 6-4

▌ THEOREM 6.5   The segments joining the midpoints of the opposite sides of any quadrilateral bisect each other.

℗ roof   Because these two segments are, in fact, the diagonals of the parallelogram formed by joining the midpoints of the consecutive sides of the quadrilateral, they bisect each other. ●

In Figure 6-5, points P, Q, R, and S are the midpoints of the sides of quadrilateral $ABCD$. The centerpoint G is determined by the intersection of $\overline{PR}$ and $\overline{QS}$. An interesting relationship exists between the segments $\overline{PR}$ and $\overline{QS}$ and the segment $\overline{MN}$ joining the midpoints M and N of the diagonals of quadrilateral $ABCD$. We state this relationship as the next theorem.

INTERACTIVE 6-5

Drag points *A*, *B*, *C*, and *D* to change the shape of the quadrilateral and see that the relationship is always true.

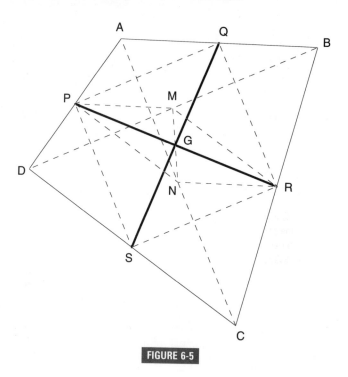

FIGURE 6-5

▌ THEOREM 6.6   The segment joining the midpoints of the diagonals of a quadrilateral is bisected by the centerpoint.

℗ roof   In Figure 6-5, M is the midpoint of $\overline{BD}$ and N is the midpoint of $\overline{AC}$. Points P, Q, R, and S are the midpoints of the sides of quadrilateral $ABCD$.

In $\triangle ADC$, $\overline{PN}$ is a midline; therefore:

$$\overline{PN} \,\|\, \overline{DC} \quad \text{and} \quad PN = \frac{1}{2}(DC)$$

In $\triangle BDC$, $\overline{MR}$ is a midline; therefore:

$$\overline{MR} \parallel \overline{DC} \quad \text{and} \quad MR = \frac{1}{2}(DC)$$

Thus $\overline{PN} \parallel \overline{MR}$ and $PN = MR$. It follows that quadrilateral $PMRN$ is a parallelogram.

The diagonals of this parallelogram bisect each other, so $\overline{MN}$ and $\overline{PR}$ share a common midpoint, G, which was earlier established as the centerpoint of the quadrilateral. ●

While we are on the topic of parallelograms, the next theorem not only presents a rather interesting relationship but together with the foregoing discussion also allows us to propose another interesting property about quadrilaterals.

THEOREM 6.7   The sum of the squares of the lengths of the sides of a parallelogram equals the sum of the squares of the lengths of the diagonals.

P roof   In the proof of Stewart's theorem ([II] and [IV]), we established the following relationships (see Figure 6-6), applied to parallelogram $ABCD$ with $\overline{BF} \perp \overline{AFEC}$.

INTERACTIVE 6-6

Drag points *A*, *B*, *C*, and *D* to change the shape of the parallelogram and see that the equation is always true.

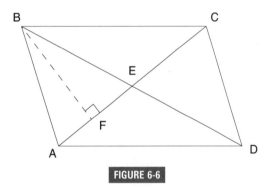

FIGURE 6-6

For $\triangle ABE$:

$$(AB)^2 = (BE)^2 + (AE)^2 - 2(AE)(FE) \tag{I}$$

For $\triangle EBC$:

$$(BC)^2 = (BE)^2 + (EC)^2 + 2(EC)(FE) \tag{II}$$

Because the diagonals of quadrilateral $ABCD$ bisect each other, $AE = EC$. Therefore, by adding equations (I) and (II), we get:

$$(AB)^2 + (BC)^2 = 2(BE)^2 + 2(AE)^2 \tag{III}$$

Similarly, in $\triangle CAD$:

$$(CD)^2 + (DA)^2 = 2(DE)^2 + 2(CE)^2 \qquad \text{(IV)}$$

Adding (III) and (IV), we get:

$$(AB)^2 + (BC)^2 + (CD)^2 + (DA)^2 = 2(BE)^2 + 2(AE)^2 + 2(DE)^2 + 2(CE)^2$$

Because $AE = EC$ and $BE = ED$, we have:

$$(AB)^2 + (BC)^2 + (CD)^2 + (DA)^2 = 4(BE)^2 + 4(AE)^2$$
$$= (2BE)^2 + (2AE)^2$$
$$= (BD)^2 + (AC)^2 \; \bullet$$

We now combine Theorems 6.1 and 6.7 to get the following theorem.

THEOREM 6.8   The sum of the squares of the lengths of the diagonals of any quadrilateral equals twice the sum of the squares of the lengths of the two segments joining the midpoints of the opposite sides of the quadrilateral.

Proof   In the proof of Theorem 6.1, we established that $PQ = \frac{1}{2}(DB)$ and $SR = \frac{1}{2}(DB)$.

INTERACTIVE 6-7

Drag points *A, B, C,* and *D* to change the shape of the quadrilateral and see that the equation is always true.

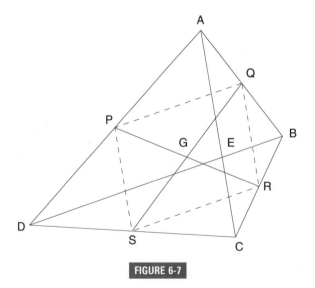

FIGURE 6-7

This gives us:

$$(PQ)^2 = \frac{1}{4}(DB)^2 \quad \text{and} \quad (SR)^2 = \frac{1}{4}(DB)^2 \qquad \text{(I)}$$

Similarly, $QR = \frac{1}{2}(AC)$ and $PS = \frac{1}{2}(AC)$. This gives us:

$$(QR)^2 = \frac{1}{4}(AC)^2 \text{ and } (PS)^2 = \frac{1}{4}(AC)^2 \qquad \text{(II)}$$

Applying Theorem 6.7 to parallelogram $PQRS$ (Figure 6-7) gives us:

$$(PQ)^2 + (SR)^2 + (QR)^2 + (PS)^2 = (PR)^2 + (QS)^2 \qquad \text{(III)}$$

Making the appropriate substitutions of (I) and (II) into (III) gives us:

$$\frac{1}{4}(DB)^2 + \frac{1}{4}(DB)^2 + \frac{1}{4}(AC)^2 + \frac{1}{4}(AC)^2 = (PR)^2 + (QS)^2$$
$$\frac{1}{2}(DB)^2 + \frac{1}{2}(AC)^2 = (PR)^2 + (QS)^2$$
$$(DB)^2 + (AC)^2 = 2[(PR)^2 + (QS)^2] \; \bullet$$

CYCLIC QUADRILATERALS

You are probably familiar with the famous formula of Heron of Alexandria for finding the area of any triangle given only the lengths of its three sides. This formula is:

$$\text{area of a triangle} = \sqrt{s(s-a)(s-b)(s-c)}$$

where a, b, and c are the lengths of the sides and $s = \dfrac{a+b+c}{2}$

It is natural to try to extend this formula to quadrilaterals. One common way is to consider the triangle as a quadrilateral with a side of zero length. Such an extension is credited to Brahmagupta,* an Indian mathematician who lived in the early part of the seventh century. He used the following formula to find the area of a *cyclic quadrilateral* (i.e., a quadrilateral that may be inscribed in a circle) with sides of length a, b, c, and d, where s is the semiperimeter:

$$\text{area} = \sqrt{(s-a)(s-b)(s-c)(s-d)}$$

---

* In 628, Brahmagupta (born 598) wrote *Brahma-sphuta-siddhānta* ("The Revised System of Brahma"), with the twelfth and thirteenth chapters devoted to mathematics.

INTERACTIVE 6-8

Drag points *A, B, C,* and *D* to change the shape of the quadrilateral and see that Brahmagupta's formula is always true.

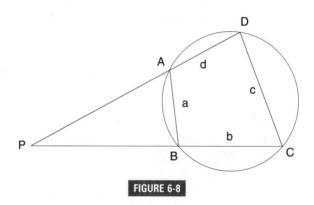

FIGURE 6-8

Note that Brahmagupta considered Heron's formula as treating the triangle as a quadrilateral with $d = 0$.

P roof **(Brahmagupta's formula)** First consider the case in which quadrilateral *ABCD* is a rectangle with $a = c$ and $b = d$. Assuming Brahmagupta's formula to be true, we have:

$$\text{area of rectangle } ABCD = \sqrt{(s - a)(s - b)(s - c)(s - d)}$$
$$= \sqrt{(a + b - a)(a + b - b)(a + b - a)(a + b - b)}$$
$$= \sqrt{a^2 b^2}$$
$$= ab$$

This is the area of the rectangle as found by the usual methods.

Now consider any nonrectangular cyclic quadrilateral *ABCD* (see Figure 6-8). Extend $\overline{DA}$ and $\overline{CB}$ to meet at point *P*, forming $\triangle DCP$.

Let $PC = x$ and $PD = y$. By Heron's formula:

$$\text{area } \triangle DCP = \frac{1}{4}\sqrt{(x + y + c)(y - x + c)(x + y - c)(x - y + c)} \quad \text{(I)}$$

Because $\angle CDA$ is supplementary to $\angle CBA$ and because $\angle ABP$ is also supplementary to $\angle CBA$, $\angle CDA \cong \angle ABP$. Thus:

$$\triangle BAP \sim \triangle DCP \quad \text{(II)}$$

From (II), we get:

$$\frac{\text{area } \triangle BAP}{\text{area } \triangle DCP} = \frac{a^2}{c^2}$$

$$\frac{\text{area } \triangle DCP}{\text{area } \triangle DCP} - \frac{\text{area } \triangle BAP}{\text{area } \triangle DCP} = \frac{c^2}{c^2} - \frac{a^2}{c^2}$$

$$\frac{\text{area } \triangle DCP - \text{area } \triangle BAP}{\text{area } \triangle DCP} = \frac{\text{area } ABCD}{\text{area } \triangle DCP} = \frac{c^2 - a^2}{c^2} \quad \text{(III)}$$

From (II), we also get:

$$\frac{x}{c} = \frac{y - d}{a} \tag{IV}$$

$$\frac{y}{c} = \frac{x - b}{a} \tag{V}$$

Adding (IV) and (V), we get:

$$\frac{x + y}{c} = \frac{x + y - b - d}{a}$$

$$x + y = \frac{c}{c - a}(b + d)$$

$$x + y + c = \frac{c}{c - a}(b + c + d - a) \tag{VI}$$

The following relationships are found by using similar methods:

$$y - x + c = \frac{c}{c + a}(a + c + d - b) \tag{VII}$$

$$x + y - c = \frac{c}{c - a}(a + b + d - c) \tag{VIII}$$

$$x - y + c = \frac{c}{c + a}(a + b + c - d) \tag{IX}$$

Now substitute (VI), (VII), (VIII), and (IX) into (I). Then:

area $\triangle DCP$

$$= \frac{c^2}{4(c^2 - a^2)}\sqrt{(b + c + d - a)(a + c + d - b)(a + b + d - c)(a + b + c - d)}$$

area $\triangle DCP$

$$= \frac{c^2}{c^2 - a^2} \cdot \frac{\sqrt{(b + c + d - a)(a + c + d - b)(a + b + d - c)(a + b + c - d)}}{4}$$

$$= \frac{c^2}{c^2 - a^2} \cdot \sqrt{\frac{(b + c + d - a)}{2}\frac{(a + c + d - b)}{2}\frac{(a + b + d - c)}{2}\frac{(a + b + c - d)}{2}}$$

$$= \frac{c^2}{c^2 - a^2} \cdot \sqrt{\frac{(a + b + c + d - 2a)}{2}\frac{(a + b + c + d - 2b)}{2}\frac{(a + b + c + d - 2c)}{2}\frac{(a + b + c + d - 2d)}{2}}$$

Because $s = \dfrac{a + b + c + d}{2}$, we get:

$$\text{area }\triangle DCP = \frac{c^2}{c^2 - a^2} \cdot \sqrt{(s - a)(s - b)(s - c)(s - d)}$$

Rewrite (III) as:

$$\text{area } \triangle DCP = \frac{c^2}{c^2 - a^2} \, (\text{area } ABCD)$$

Thus the area of cyclic quadrilateral $ABCD = \sqrt{(s-a)(s-b)(s-c)(s-d)}.$ ●

An interesting extension of Brahmagupta's formula to the general quadrilateral is given here without proof:

$$\text{area of any (convex) quadrilateral}$$
$$= \sqrt{(s-a)(s-b)(s-c)(s-d) - abcd \cdot \cos^2\left(\frac{\alpha + \gamma}{2}\right)}$$

where a, b, c, and d are the lengths of the sides, $s = \dfrac{a+b+c+d}{2}$, and α and γ are the measures of a pair of opposite angles of the quadrilateral.

This formula shows that of all quadrilaterals that can be formed from four given side lengths, the one with the maximum area is the cyclic quadrilateral. The maximum area is achieved when $abcd \cdot \cos^2\left(\dfrac{\alpha + \gamma}{2}\right) = 0$, which occurs when $\alpha + \gamma = 180°$, which is true only in cyclic quadrilaterals.

There are many interesting theorems about cyclic quadrilaterals. Before considering them, the reader is advised to recall the methods of proving that a quadrilateral is cyclic (see page 7).

Brahmagupta also found that for a cyclic quadrilateral of consecutive side lengths a, b, c, and d, where m and n are the lengths of the diagonals, the following relationships hold true:

$$m^2 = \frac{(ab + cd)(ac + bd)}{ad + bc}$$
$$n^2 = \frac{(ac + bd)(ad + bc)}{ab + cd}$$

Another interesting theorem on cyclic quadrilaterals attributed to Brahmagupta follows.

▌THEOREM 6.9   In a cyclic quadrilateral with perpendicular diagonals, the line through the point of intersection of the diagonals and perpendicular to a side of the quadrilateral bisects the opposite side.

---

❷roof   Diagonals $\overline{AC}$ and $\overline{BD}$ of cyclic quadrilateral $ABCD$ are perpendicular at point of intersection, G, and $\overleftrightarrow{GE} \perp \overleftrightarrow{AED}$ (see Figure 6-9). We want to prove that $\overleftrightarrow{GE}$ bisects $\overline{BC}$ at point P.

In right triangle AEG, $\angle 5$ is complementary to $\angle 1$ and $\angle 2$ is complementary to $\angle 1$. Therefore $\angle 5 \cong \angle 2$. Because $\angle 2 \cong \angle 4$, we have $\angle 5 \cong \angle 4$.

INTERACTIVE 6-9

Drag points *A, B, C,* and *D* to change the shape of the quadrilateral and see that point *P* always bisects $\overline{CB}$.

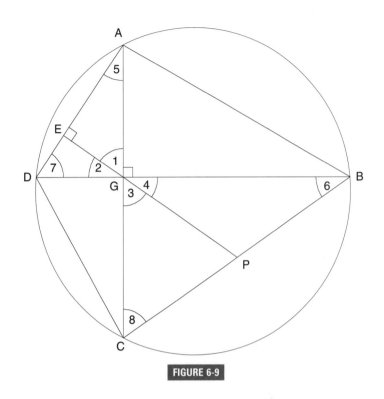

FIGURE 6-9

Because $\angle 5$ and $\angle 6$ are equal in measure to $\frac{1}{2} m\widehat{DC}$, they are congruent. Therefore $\angle 4 \cong \angle 6$ and $BP = GP$.

Similarly, because $\angle 7 \cong \angle 3$ and $\angle 7 \cong \angle 8$, we have $GP = PC$. Thus $CP = PB$. ●

An interesting way to generate a cyclic quadrilateral is provided by the next theorem.

THEOREM 6.10   If from each pair of adjacent angles of any quadrilateral the angle bisectors are drawn, the segments connecting their points of intersection form a cyclic quadrilateral.

℗roof   In Figure 6-10, the angle bisectors of quadrilateral *ABCD* meet to determine quadrilateral *EFGH*. We will prove this latter quadrilateral to be cyclic.

$$m\angle BAD + m\angle ADC + m\angle DCB + m\angle CBA = 360°$$

Therefore:

$$\frac{1}{2} m\angle BAD + \frac{1}{2} m\angle ADC + \frac{1}{2} m\angle DCB + \frac{1}{2} m\angle CBA = \frac{1}{2}(360°) = 180°$$

Drag points *A, B, C,* and *D* to change the shape of the quadrilateral and see that *EFGH* is always cyclic.

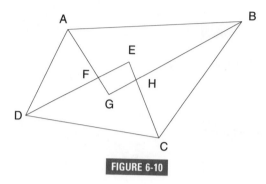

FIGURE 6-10

Substituting, we have:

$$m\angle EDC + m\angle ECD + m\angle GAB + m\angle ABG = 180°$$ (I)

Consider $\triangle ABG$ and $\triangle DEC$:

$$m\angle EDC + m\angle ECD + m\angle GAB + m\angle ABG + m\angle AGB \\ + m\angle DEC = 2(180°)$$ (II)

Subtracting (I) from (II), we find that:

$$m\angle AGB + m\angle DEC = 180°$$

Because one pair of opposite angles of quadrilateral *EFGH* are supplementary, the other pair must also be supplementary, and hence quadrilateral *EFGH* is cyclic. ●

PTOLEMY'S THEOREM

Perhaps the most famous theorem involving cyclic quadrilaterals is that attributed to Claudius Ptolemaeus of Alexandria (popularly known as Ptolemy). In his major astronomical work, the *Almagest** (ca. A.D. 150), he stated this theorem on cyclic quadrilaterals.

THEOREM 6.11   **(Ptolemy's theorem)**   The product of the lengths of the diagonals of a cyclic quadrilateral equals the sum of the products of the lengths of the pairs of opposite sides.

* The Greek title, *Syntaxis Mathematica,* means "mathematical (or astronomical) compilation." The Arabic title, *Almagest,* is a renaming meaning "great collection (or compilation)." The book is a manual of all the mathematical astronomy that the ancients knew up to that time. Book I of the thirteen books that comprise this monumental work contains the theorem that now bears Ptolemy's name.

We provide two methods for proving Ptolemy's theorem. The second method incorporates the proof of the converse as well, which we state in Theorem 6.12.

℗roof I

INTERACTIVE 6-11

Drag points *A, B, C,* and *D* to change the shape of the quadrilateral and see that the equation is always true.

In Figure 6-11, quadrilateral *ABCD* is inscribed in circle *O*. A line is drawn through point *A* to meet $\overleftrightarrow{GE}$ at point *P* so that:

$$m\angle BAC = m\angle DAP \qquad \text{(I)}$$

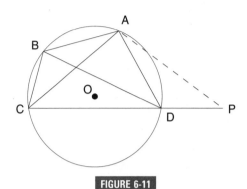

FIGURE 6-11

Because quadrilateral *ABCD* is cyclic, $\angle ABC$ is supplementary to $\angle ADC$. However, $\angle ADP$ is also supplementary to $\angle ADC$. Therefore:

$$m\angle ABC = m\angle ADP \qquad \text{(II)}$$

Thus:

$$\triangle BAC \sim \triangle DAP \text{ (AA)} \qquad \text{(III)}$$

$$\frac{AB}{AD} = \frac{BC}{DP} \quad \text{or} \quad DP = \frac{(AD)(BC)}{AB} \qquad \text{(IV)}$$

From (I), $m\angle BAD = m\angle CAP$; from (III), $\dfrac{AB}{AD} = \dfrac{AC}{AP}$. Therefore:

$$\triangle ABD \sim \triangle ACP \text{ (SAS)}$$

$$\frac{BD}{CP} = \frac{AB}{AC} \quad \text{or} \quad CP = \frac{(AC)(BD)}{AB} \qquad \text{(V)}$$

We know that:

$$CP = CD + DP \qquad \text{(VI)}$$

Substituting (IV) and (V) into (VI), we get:

$$\frac{(AC)(BD)}{AB} = CD + \frac{(AD)(BC)}{AB}$$

Thus $(AC)(BD) = (AB)(CD) + (AD)(BC)$. ●

℗roof II

In quadrilateral *ABCD* (Figure 6-12), draw $\triangle DAP$ on side $\overline{AD}$ similar to $\triangle CAB$. Thus:

$$\frac{AB}{AP} = \frac{AC}{AD} = \frac{BC}{PD} \qquad \text{(I)}$$

$$(AC)(PD) = (AD)(BC) \qquad \text{(II)}$$

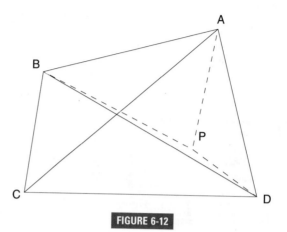

FIGURE 6-12

Because $m\angle BAC = m\angle PAD$, $m\angle BAP = m\angle CAD$. Therefore, from (I):

$$\triangle BAP \sim \triangle CAD \text{ (SAS)} \Rightarrow \frac{AB}{AC} = \frac{BP}{CD}$$

Thus:

$$(AC)(BP) = (AB)(CD) \qquad \text{(III)}$$

Adding (II) and (III), we have:

$$(AC)(BP + PD) = (AD)(BC) + (AB)(CD) \qquad \text{(IV)}$$

Now $BP + PD > BD$ (triangle inequality), unless P is on $\overline{BD}$. However, point P will be on $\overline{BD}$ if and only if $m\angle ADP = m\angle ADB$. But we already know that $m\angle ADP = m\angle ACB$ (similar triangles). If quadrilateral $ABCD$ was cyclic, then $m\angle ADB$ would equal $m\angle ACB$ and $m\angle ADB$ would equal $m\angle ADP$. Therefore we can state that if and only if quadrilateral $ABCD$ is cyclic, point P lies on $\overline{BD}$. This tells us that:

$$BP + PD = BD \qquad \text{(V)}$$

Substituting (V) into (IV), we get:

$$(AC)(BD) = (AD)(BC) + (AB)(CD) \ \bullet$$

Notice that we have proved both Ptolemy's theorem and its converse, which we now state as our next theorem.

THEOREM 6.12 **(The converse of Ptolemy's theorem)** If the product of the lengths of the diagonals of a quadrilateral equals the sum of the products of the lengths of the pairs of opposite sides, then the quadrilateral is cyclic.

Proof   Assume quadrilateral $ABCD$ is not cyclic (see Figure 6-11). If $\overline{CDP}$, then $m\angle ADP \neq m\angle ABC$. If points C, D, and P are not collinear, then it is possible to have $m\angle ADP = m\angle ABC$. However, then $CP < CD + DP$, and from (IV) and (V) in Proof I of Ptolemy's theorem:

$$(AC)(BD) < (AB)(CD) + (AD)(BC)$$

But this contradicts the given information that

$$(AC)(BD) = (AB)(CD) + (AD)(BC)$$

Therefore quadrilateral $ABCD$ is cyclic. ●

We now consider an extension of Ptolemy's theorem.

THEOREM 6.13 Consider a noncyclic quadrilateral $ABCD$. If we let $a = AD$, $b = BD$, $c = CD$, $a' = BC$, $b' = AC$, and $c' = AB$, then the sum of the lengths of any two of segments aa', bb', and cc' is greater than the length of the third (see Figure 6-13).

Proof* Construct A' on $\overrightarrow{DA}$ such that $DA' = bc$.

Construct B' on $\overrightarrow{DB}$ such that $DB' = ac$.

Construct C' on $\overrightarrow{DC}$ such that $DC' = ab$.

We notice that $\triangle DAB \sim \triangle DA'B'$ because they have a common $\angle ADB$ and the adjacent sides are proportional. That is:

$$\frac{DA'}{DB} = \frac{bc}{b} = c \quad \text{and} \quad \frac{DB'}{DA} = \frac{ac}{a} = c \Rightarrow \frac{DA'}{DB} = \frac{DB'}{DA}$$

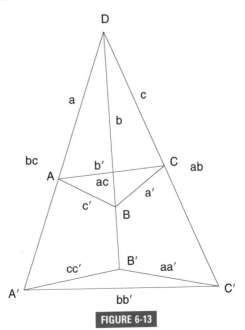

FIGURE 6-13

---

* This proof was developed by Professor Harry W. Appelgate of The City College of The City University of New York.

It follows that $c = \dfrac{A'B'}{AB} = \dfrac{x}{c'}$, which gives us $A'B' = cc'$. Similarly, $B'C' = aa'$ and $A'C' = bb'$. From $\triangle A'B'C'$, we see (from the triangle inequality) that:

$$aa' + bb' > cc'$$
$$aa' + cc' > bb'$$
$$bb' + cc' > aa' \quad \bullet$$

For the situation in Theorem 6.13, what happens when $aa' + cc' = bb'$?

APPLICATIONS OF PTOLEMY'S THEOREM

This section presents some direct consequences of Ptolemy's theorem.

Application 1   If any circle passing through vertex A of parallelogram $ABCD$ intersects sides $\overline{AB}$ and $\overline{AD}$ at points P and R, respectively, and intersects diagonal $\overline{AC}$ at point Q, prove that $(AQ)(AC) = (AP)(AB) + (AR)(AD)$. $\bullet$

Proof   Draw $\overline{RQ}$, $\overline{QP}$, and $\overline{RP}$, as in Figure 6-14. $m\angle 4 = m\angle 2$. Similarly, $m\angle 1 = m\angle 3$. Because $m\angle 5 = m\angle 3$, $m\angle 1 = m\angle 5$.

Therefore $\triangle RQP \sim \triangle ABC$ (AA), and because $\triangle ABC \cong \triangle CDA$, we have:

$$\triangle RQP \sim \triangle ABC \sim \triangle CDA$$

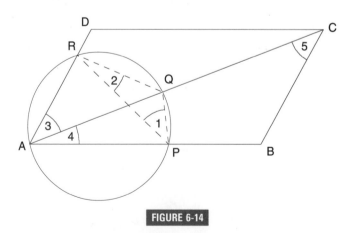

FIGURE 6-14

Then:

$$\frac{AC}{RP} = \frac{AB}{RQ} = \frac{AD}{PQ} \tag{I}$$

By Ptolemy's theorem, in quadrilateral $RQPA$:

$$(AQ)(RP) = (RQ)(AP) + (PQ)(AR) \tag{II}$$

Multiplying each of the three equal ratios in (I) by one member of (II) gives us:

$$(AQ)(RP)\left(\frac{AC}{RP}\right) = (RQ)(AP)\left(\frac{AB}{RQ}\right) + (PQ)(AR)\left(\frac{AD}{PQ}\right)$$

Thus $(AQ)(AC) = (AP)(AB) + (AR)(AD)$. ●

Ⓐpplication 2   Express the ratio of the lengths of the diagonals of a cyclic quadrilateral in terms of the lengths of the sides. ●

Ⓢolution   On the circumcircle of quadrilateral $ABCD$, choose points P and Q so that $PA = DC$ and $QD = AB$, as in Figure 6-15.
Applying Ptolemy's theorem to quadrilateral $ABCP$ gives us:

$$(AC)(PB) = (AB)(PC) + (BC)(PA) \tag{I}$$

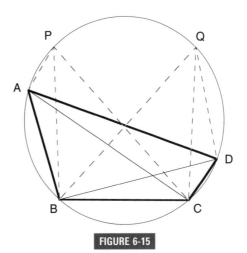

FIGURE 6-15

Similarly, applying Ptolemy's theorem to quadrilateral $BCDQ$ gives us:

$$(BD)(QC) = (DC)(QB) + (BC)(QD) \tag{II}$$

Because $PA + AB = DC + QD$, we have $m\,\widehat{PAB} = m\,\widehat{QDC}$ and $PB = QC$. Similarly, because $m\,\widehat{PBC} = m\,\widehat{DBA}$, we have $PC = AD$, and because $m\,\widehat{QCB} = m\,\widehat{ACD}$, we have $QB = AD$.

Finally, dividing (I) by (II) and substituting for all terms containing Q and P, we get:

$$\frac{AC}{BD} = \frac{(AB)(AD) + (BC)(DC)}{(DC)(AD) + (BC)(AB)} \bullet$$

Application 3   A point P is chosen inside parallelogram $ABCD$ such that $\angle APB$ is supplementary to $\angle CPD$ (Figure 6-16). Prove that $(AB)(AD) = (BP)(DP) + (AP)(CP)$. $\bullet$

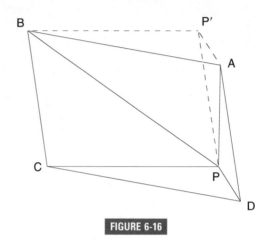

FIGURE 6-16

Proof   On side $\overline{AB}$ of parallelogram $ABCD$, draw $\triangle AP'B \cong \triangle DPC$ so that:

$$DP = AP' \quad \text{and} \quad CP = BP' \tag{I}$$

Because $\angle APB$ is supplementary to $\angle CPD$ and $m\angle BP'A = m\angle CPD$, $\angle APB$ is supplementary to $\angle BP'A$. Therefore quadrilateral $BP'AP$ is cyclic. Applying Ptolemy's theorem to cyclic quadrilateral $BP'AP$, we get:

$$(AB)(P'P) = (BP)(AP') + (AP)(BP')$$

From (I):

$$(AB)(P'P) = (BP)(DP) + (AP)(CP) \tag{II}$$

Because $m\angle BAP' = m\angle CDP$ and $\overline{CD} \parallel \overline{AB}$, we have $\overline{PD} \parallel \overline{P'A}$. Therefore $PDAP'$ is a parallelogram and $P'P = AD$.

Thus, from (II):

$$(AB)(AD) = (BP)(DP) + (AP)(CP) \bullet$$

The next five applications develop a rather nice pattern about regular polygons.

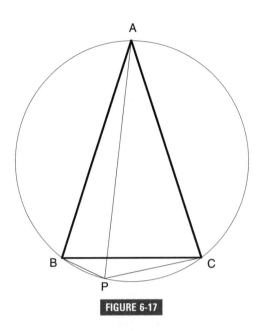

FIGURE 6-17

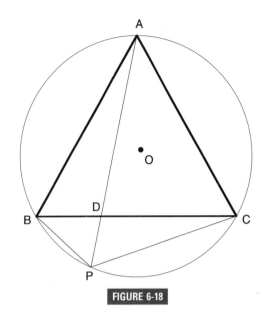

FIGURE 6-18

Application 4 If isosceles triangle ABC ($AB = AC$) is inscribed in a circle and point P is on $\overset{\frown}{BC}$, prove that $\dfrac{PA}{PB + PC} = \dfrac{AC}{BC}$, a constant for the given triangle. ●

Proof Applying Ptolemy's theorem to cyclic quadrilateral $ABPC$ (Figure 6-17), we get:

$$(PA)(BC) = (PB)(AC) + (PC)(AB)$$

Because $AB = AC$:

$$(PA)(BC) = AC(PB + PC) \quad \text{and} \quad \frac{PA}{PB + PC} = \frac{AC}{BC} \quad ●$$

Application 5 If equilateral triangle ABC is inscribed in a circle and point P is on $\overset{\frown}{BC}$, prove that $PA = PB + PC$. ●

Proof Because quadrilateral $ABPC$ is cyclic (Figure 6-18), we can apply Ptolemy's theorem:

$$(PA)(BC) = (PB)(AC) + (PC)(AB) \tag{I}$$

However, because $\triangle ABC$ is equilateral, $BC = AC = AB$. Therefore, from (I):

$$PA = PB + PC \quad ●$$

Application 6   If square $ABCD$ is inscribed in a circle and point P is on $\overset{\frown}{BC}$, prove that $\dfrac{PA + PC}{PB + PD} = \dfrac{PD}{PA}$. ●

Proof   In Figure 6-19, consider isosceles triangle ABD ($AB = AD$). Using the results of Application 4, we have:

$$\frac{PA}{PB + PD} = \frac{AD}{DB} \tag{I}$$

Similarly, in isosceles triangle ADC:

$$\frac{PD}{PA + PC} = \frac{DC}{AC} \tag{II}$$

Because $AD = DC$ and $DB = AC$, we have:

$$\frac{AD}{DB} = \frac{DC}{AC} \tag{III}$$

From (I), (II), and (III):

$$\frac{PA}{PB + PD} = \frac{PD}{PA + PC} \quad \text{or} \quad \frac{PA + PC}{PB + PD} = \frac{PD}{PA} ●$$

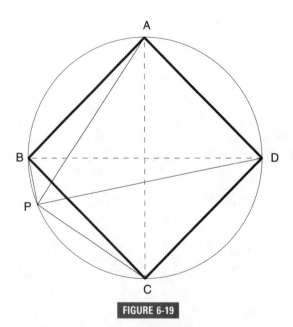

FIGURE 6-19

Application 7   If regular pentagon $ABCDE$ is inscribed in a circle and point P is on $\overset{\frown}{BC}$, prove that $PA + PD = PB + PC + PE$. ●

Proof   In quadrilateral $ABPC$ (see Figure 6-20), by Ptolemy's theorem:

$$(PA)(BC) = (BA)(PC) + (PB)(AC) \tag{I}$$

In quadrilateral $BPCD$:

$$(PD)(BC) = (CD)(PB) + (PC)(BD) \tag{II}$$

Because $BA = CD$ and $AC = BD$, by adding (I) and (II) we obtain:

$$BC(PA + PD) = BA(PB + PC) + AC(PB + PC) \tag{III}$$

However, because $\triangle BEC$ is isosceles, based on Application 4 we have:

$$\frac{CE}{BC} = \frac{PE}{PB + PC} \quad \text{or} \quad \frac{(PE)(BC)}{(PB + PC)} = CE = AC \tag{IV}$$

Substituting (IV) into (III) gives us:

$$BC(PA + PD) = BA(PB + PC) + \frac{(PE)(BC)}{(PB + PC)}(PB + PC)$$

But $BC = BA$. Therefore $PA + PD = PB + PC + PE$. ●

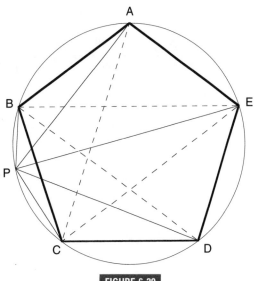

FIGURE 6-20

Ⓐpplication 8   If regular hexagon *ABCDEF* is inscribed in a circle and point *P* is on $\overset{\frown}{BC}$, prove that *PE* + *PF* = *PA* + *PB* + *PC* + *PD*. ●

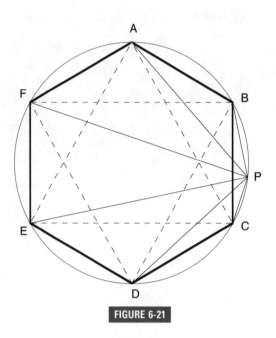

FIGURE 6-21

Ⓟroof   Lines are drawn between points *A*, *E*, and *C* to make equilateral triangle *AEC* (Figure 6-21). Using the results of Application 5, we have:

$$PE = PA + PC \tag{I}$$

In the same way, in equilateral triangle *BFD*:

$$PF = PB + PD \tag{II}$$

Adding (I) and (II), we get:

$$PE + PF = PA + PB + PC + PD \; ●$$

Although the following problem can be solved by other means, the solution that uses Ptolemy's theorem is rather nice.

Application 9   A triangle inscribed in a circle of radius 5 has two sides of length 5 and 6. Find the length of the third side of the triangle. ●

Solution   In Figure 6-22, we notice that there are two cases to consider in this problem. Both $\triangle ABC$ and $\triangle ABC'$ are inscribed in circle O, with $AB = 5$ and $AC = AC' = 6$. We are to find BC and BC'.

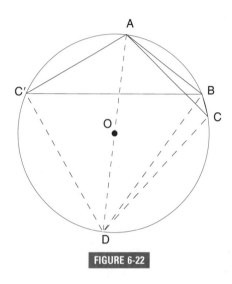

FIGURE 6-22

Draw diameter $\overline{AOD}$, which measures 10, and draw $\overline{DC}$, $\overline{DB}$, and $\overline{DC'}$. Then:

$$m\angle AC'D = m\angle ACD = m\angle ABD = 90°$$

Consider the case where $\angle A$ in $\triangle ABC$ is acute. In right triangle ACD, $DC = 8$, and in right triangle ABD, $BD = 5\sqrt{3}$. Applying Ptolemy's theorem to quadrilateral $ABCD$:

$$(AC)(BD) = (AB)(DC) + (AD)(BC)$$
$$(6)(5\sqrt{3}) = (5)(8) + (10)(BC) \quad \text{or} \quad BC = 3\sqrt{3} - 4$$

Now consider the case where $\angle A$ is obtuse, as in $\triangle ABC'$. In right triangle $AC'D$, $DC' = 8$. Applying Ptolemy's theorem to quadrilateral $ABDC'$:

$$(AC')(BD) + (AB)(DC') = (AD)(BC)$$
$$(6)(5\sqrt{3}) = (5)(8) = (10)(BC') \quad \text{or} \quad BC' = 3\sqrt{3} + 4 ●$$

We began our study of quadrilaterals by investigating properties of the general quadrilateral. This led to a consideration of cyclic quadrilaterals, quadrilaterals whose areas are a maximum for the given side lengths. The cyclic quadrilaterals are rich in interesting properties. Brahmagupta's formula and Ptolemy's theorem gave evidence of this. It is up to the reader to continue investigating the properties of various other kinds of quadrilaterals. The field is boundless, and the resulting pleasure a certainty.

E X E R C I S E S

1. What type of quadrilateral is formed by joining the midpoints of consecutive sides of:
    a. a nonisosceles trapezoid
    b. an isosceles trapezoid
    Prove your answers.

2. If two noncongruent isosceles triangles share a common base and have no part of their interior regions in common, determine the type of quadrilateral formed by joining the midpoints of consecutive sides of the quadrilateral (formed by the two isosceles triangles).

3. Is the converse of Theorem 6.1 true? Prove your answer.

4. Prove that the perimeter of the quadrilateral formed by joining the midpoints of consecutive sides of a given quadrilateral equals the sum of the lengths of the diagonals of the given quadrilateral.

5. Prove that the area of the quadrilateral formed by joining the midpoints of consecutive sides of a given quadrilateral equals one-half the area of the given quadrilateral.

6. Prove that the sum of the squares of the lengths of the sides of a quadrilateral equals the sum of the squares of the lengths of the diagonals plus four times the square of the length of the segment joining the midpoints of the diagonals.

7. Find the area of a triangle whose sides have lengths 13, 14, and 15.

8. Find the area of a cyclic quadrilateral whose sides have lengths 9, 10, 10, and 21.

9. Find the area of a cyclic quadrilateral whose sides have lengths 7, 15, 20, and 24.

10. A line, $\overleftrightarrow{PQ}$, parallel to base $\overline{BC}$ of $\triangle ABC$ intersects $\overline{AB}$ and $\overline{AC}$ at points P and Q, respectively (Figure 6-23). The circle passing through point P and tangent to $\overline{AC}$ at point Q intersects $\overline{AB}$ again at point R. Prove that points R, Q, C, and B lie on a circle.

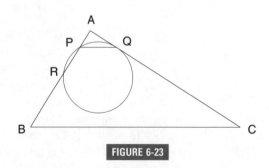

FIGURE 6-23

11. Prove that the lines from the midpoints of the sides of a cyclic quadrilateral and perpendicular to the respective opposite sides are concurrent.

12. To what familiar result does Ptolemy's theorem lead when the cyclic quadrilateral is a rectangle? Prove your result.

13. E is a point on side $\overline{AD}$ of rectangle $ABCD$ so that $DE = 6$, while $DA = 8$ and $DC = 6$ (see Figure 6-24). If $\overline{CE}$ is extended to meet the circumcircle of the rectangle at point F, find the length of $\overline{DF}$. Also find the length of $\overline{FB}$.

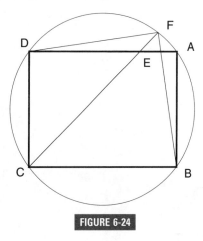

FIGURE 6-24

14. A line drawn from vertex A of equilateral triangle ABC meets $\overline{BC}$ at point D and the circumcircle at point P (see Figure 6-25). Prove that $\dfrac{1}{PD} = \dfrac{1}{PB} + \dfrac{1}{PC}$.

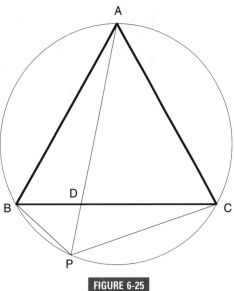

FIGURE 6-25

15. Prove that a quadrilateral has perpendicular diagonals if and only if the sum of the squares of the lengths of one pair of opposite sides equals the sum of the squares of the lengths of the other pair of opposite sides.

EQUICIRCLES

POINTS OF TANGENCY

We begin this chapter in somewhat dramatic form. We recall a common theorem from elementary geometry:

THEOREM 7.0   Two tangent segments to a circle from an external point are equal in length.

Consider the following problem, whose solution uses this theorem a few times.

If the perimeter of $\triangle ABC$ (Figure 7-1) is 16, find the length of $\overline{AK_1}$. (Note: Each of the four circles I, I_1, I_2, I_3 is tangent to each of the three lines forming $\triangle ABC$.)

INTERACTIVE 7-1

Drag points A, B, and C to change the shape of the triangle and see that AK_1 is always $\frac{1}{2}$ the perimeter.

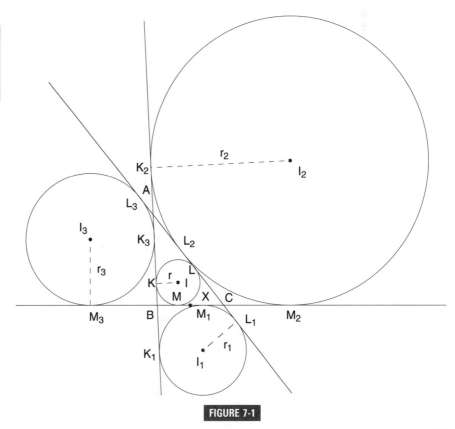

FIGURE 7-1

Applying Theorem 7.0 to this figure, we get:

$$BK_1 = BM_1 \quad \text{and} \quad CL_1 = CM_1$$

The perimeter of $\triangle ABC = AB + BC + AC = AB + (BM_1 + CM_1) + AC$. By substitution:

$$\text{perimeter } \triangle ABC = AB + BK_1 + CL_1 + AC$$
$$= AK_1 + AL_1$$

However, $AK_1 = AL_1$ because these two segments are tangent segments from the same external point to the same circle (Theorem 7.0). Therefore:

$$AK_1 = \frac{1}{2}(\text{perimeter } \triangle ABC) = 8$$

This solution exhibits just one of the many interesting relationships involving *equicircles*, which are the three *escribed circles*, or *excircles*, and the *inscribed circle*, or *incircle*, of a triangle. In the discussion that follows, we will investigate other interesting relationships involving the equicircles and their points of tangency. Each equicircle is tangent to *each* of the three lines containing the sides of the triangle.

Let us first state formally the relationship developed above.

THEOREM 7.1   The segment of a line, containing a side of a triangle, from the vertex of a triangle to the point of contact of the opposite excircle has a length equal to half the length of the perimeter of the triangle.

By letting s represent the semiperimeter of $\triangle ABC$, we can restate this theorem as:

$$AK_1 = AL_1 = s$$

Recall that $a = BC$, $b = AC$, and $c = AB$. Therefore:

$$BM_1 = BK_1 = AK_1 - AB = s - c$$
$$CM_1 = CL_1 = AL_1 - AC = s - b$$

Other, analogous relationships can easily be found.

THEOREM 7.2   The segment of a side of a triangle from the vertex to the point of tangency of an excircle has a length equal to the semiperimeter minus the length of the other adjacent side.

At this point we should briefly recall how equicircles are determined. In Chapter 2, we proved (Application 3) that the interior angle bisector of a triangle is concurrent with the two exterior angle bisectors, one at each of the other two vertices. This point of concurrency is the center of an escribed circle and is called the *excenter* of the triangle. Using the property that all points of an angle bisector are equidistant from the sides of the angle, we can easily prove that this point of concurrency is, in fact, the center of an escribed circle. From elementary geometry, you will recall that the center of the inscribed circle is the point of intersection of the interior angle bisectors of the triangle.

Despite all the work students of elementary geometry do with tangent segments from a common exterior point, they never consider the length of these segments in terms of the lengths of the sides of a triangle formed by common tangents. A proof of the next theorem fills this void.

THEOREM 7.3   The segment of a side of a triangle from a vertex to the point of tangency of the incircle has a length equal to the semiperimeter minus the length of the opposite side.

Proof   In Figure 7-1:

$$AK + AL = AB - KB + AC - LC = AB + AC - (KB + LC)$$

However, $KB = MB$ and $LC = MC$. By substitution, we get:

$$
\begin{aligned}
AK + AL &= AB + AC - (MB + MC) \\
&= AB + AC - BC \\
&= c + b - a \\
&= 2s - 2a = 2(s - a) \text{ where } s = \frac{a + b + c}{2}
\end{aligned}
$$

Because $AK = AL$:

$$AK = AL = s - a$$

In a similar manner, we get:

$$BM = BK = s - b \quad \text{and} \quad CL = CM = s - c$$

This proves our theorem. ●

Some other interesting relationships involving the points of tangency of the four equicircles of a triangle follow.

THEOREM 7.4   The length of the common internal tangent segment of the incircle and an excircle is equal to the difference of the lengths of the two sides not containing the common internal tangent.

Proof   We wish to find the length of $\overline{MM_1}$ in terms of the lengths of the sides of the triangle.

From Theorem 7.2, we have:

$$CM_1 = CL_1 = s - b$$

From Theorem 7.3, we have:

$$BM = BK = s - b$$

Now:

$$
\begin{aligned}
MM_1 &= BC - BM - CM_1 \\
&= a - (s - b) - (s - b) \\
&= a - 2(s - b) \\
&= a + 2b - 2s \text{ where } s = \frac{a + b + c}{2} \\
MM_1 &= b - c \quad \bullet
\end{aligned}
$$

Directly from this proof we can show that the midpoint of $\overline{MM_1}$ is also the midpoint of $\overline{BC}$. We begin by noting that $BM = CM_1$ and, for midpoint X of $\overline{BC}$, $BX = CX$. By subtraction, $MX = M_1X$. We state this result as our next theorem.

THEOREM 7.5   The midpoint of a side of a triangle is also the midpoint of the common internal tangent segment of the incircle and an excircle.

A natural concern for us to pursue here is what the length is of $\overline{MM_3}$, the segment joining the points of tangency of the incircle and an *adjacent* excircle, that is, the common external tangent segment of the incircle and an excircle.

This length is easily established. We have $MM_3 = CM_3 - CM$. From Theorem 7.1:

$$CM_3 = s$$

From Theorem 7.3:

$$CM = s - c$$

Therefore:

$$MM_3 = s - (s - c) \quad \text{or} \quad MM_3 = c = LL_3$$

We state this result as Theorem 7.6.

THEOREM 7.6   The length of the common external tangent segment of the incircle and an excircle equals the length of the side between these two circles.

We now consider the common external tangent segments of two excircles.

THEOREM 7.7   The length of the common external tangent segment of two excircles equals the sum of the lengths of the two sides whose lines intersect this common external tangent.

Proof   We seek to find the length of $\overline{M_2M_3}$.

$$M_2M_3 = MM_2 + MM_3$$

From Theorem 7.6, $MM_2 = b$ and $MM_3 = c$. Therefore:

$$M_2M_3 = b + c \; \bullet$$

This leaves us only to determine the length of the common internal tangents of two excircles of a triangle.

THEOREM 7.8   The length of a common internal tangent segment of two excircles of a triangle equals the length of the side opposite the vertex contained by the tangent segment.

Proof   We seek to find the length of $\overline{M_1M_2}$.

$$M_1M_2 = MM_2 - MM_1$$

From Theorem 7.6:

$$MM_2 = b$$

From Theorem 7.4:

$$MM_1 = b - c$$

By appropriate substitution, we get:

$$M_1M_2 = b - (b - c) \quad \text{or} \quad M_1M_2 = c \; \bullet$$

This completes our consideration of the segments determined by the points of tangency of the four equicircles of a triangle.

EQUIRADII

INTERACTIVE 7-2

Drag points *A, B,* and *C* to
change the shape of the triangle
and see that the equation is
always true.

We now turn our attention to the radii of the equicircles of a triangle. We refer
to these radii as *equiradii*. Consider first the radius of the incircle. We call this
the *inradius* of the triangle.

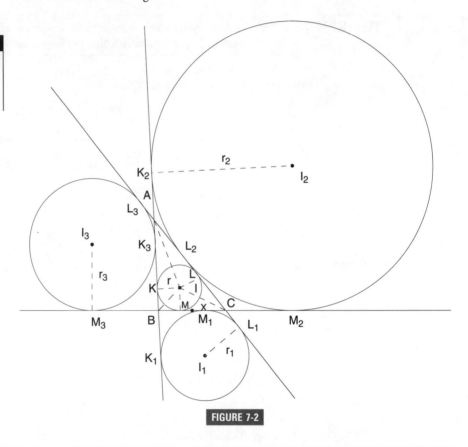

FIGURE 7-2

▍THEOREM 7.9 The length of the inradius of a triangle is equal to the ratio of the area of the
triangle to its semiperimeter.

Ⓟroof In Figure 7-2:

$$\text{area } \triangle ABC = \text{area } \triangle BCI + \text{area } \triangle ACI + \text{area } \triangle ABI$$

$$= \frac{1}{2}(IM)(BC) + \frac{1}{2}(IL)(AC) + \frac{1}{2}(IK)(AB)$$

$$= \frac{1}{2}ra + \frac{1}{2}rb + \frac{1}{2}rc = \frac{1}{2}r(a + b + c) = sr$$

Therefore:

$$r = \frac{\text{area } \triangle ABC}{s} \quad ●$$

We now examine the *exradii* (the radii of the excircles).

THEOREM 7.10   An exradius of a triangle has length equal to the ratio of the area of the triangle to the difference between the semiperimeter and the length of the side to which the excircle considered is internally tangent.

𝓟roof   In Figure 7-3:

$$\text{area } \triangle ABC = \text{area } \triangle ABI_1 + \text{area } \triangle ACI_1 - \text{area } \triangle BCI_1$$

$$= \frac{1}{2}(I_1 K_1)(AB) + \frac{1}{2}(I_1 L_1)(AC) - \frac{1}{2}(I_1 M_1)(BC)$$

$$= \frac{1}{2} r_1 c + \frac{1}{2} r_1 b - \frac{1}{2} r_1 a = \frac{1}{2} r_1 (c + b - a) = r_1(s - a)$$

Therefore:

$$r_1 = \frac{\text{area } \triangle ABC}{s - a}$$

In a similar manner, we can establish that:

$$r_2 = \frac{\text{area } \triangle ABC}{s - b} \quad \text{and} \quad r_3 = \frac{\text{area } \triangle ABC}{s - c} \quad \bullet$$

INTERACTIVE 7-3

Drag points *A*, *B*, and *C* to change the shape of the triangle and see that the equation is always true.

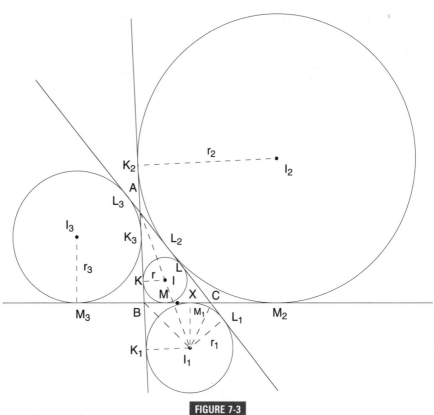

FIGURE 7-3

If we multiply the results of Theorems 7.9 and 7.10, we get:

$$r \cdot r_1 \cdot r_2 \cdot r_3 = \frac{\text{area } \triangle ABC}{s} \cdot \frac{\text{area } \triangle ABC}{s-a} \cdot \frac{\text{area } \triangle ABC}{s-b} \cdot \frac{\text{area } \triangle ABC}{s-c}$$

$$= \frac{(\text{area } \triangle ABC)^4}{s(s-a)(s-b)(s-c)}$$

This denominator reminds us of Heron's formula for finding the area of a triangle:

$$\text{area } \triangle ABC = \sqrt{s(s-a)(s-b)(s-c)}$$

Thus $(\text{area } \triangle ABC)^2 = s(s-a)(s-b)(s-c)$. By substitution:

$$r \cdot r_1 \cdot r_2 \cdot r_3 = (\text{area } \triangle ABC)^2$$

We state this formula as our next theorem.

▌THEOREM 7.11   The product of the lengths of the equiradii of a triangle equals the square of the area of the triangle.

Theorems 7.9 and 7.10 also enable us to prove the following theorem.

▌THEOREM 7.12   The reciprocal of the length of the inradius of a triangle equals the sum of the reciprocals of the lengths of the exradii of the triangle.

℗roof   From Theorem 7.10, we have:

$$\frac{1}{r_1} + \frac{1}{r_2} + \frac{1}{r_3} = \frac{s-a}{\text{area } \triangle ABC} + \frac{s-b}{\text{area } \triangle ABC} + \frac{s-c}{\text{area } \triangle ABC}$$

$$= \frac{3s - (a+b+c)}{\text{area } \triangle ABC}$$

$$= \frac{3s - 2s}{\text{area } \triangle ABC}$$

$$= \frac{s}{\text{area } \triangle ABC}$$

By Theorem 7.9, we have:

$$\frac{1}{r} = \frac{s}{\text{area } \triangle ABC}$$

Therefore:

$$\frac{1}{r_1} + \frac{1}{r_2} + \frac{1}{r_3} = \frac{1}{r} \; \bullet$$

A similar relationship exists involving altitudes h_a, h_b, and h_c of $\triangle ABC$.

THEOREM 7.13 The sum of the reciprocals of the lengths of the altitudes of a triangle equals the reciprocal of the length of the inradius.

℗roof Begin by representing the area of $\triangle ABC$ in three different ways.

$$\text{area } \triangle ABC = \frac{1}{2}ah_a = \frac{1}{2}bh_b = \frac{1}{2}ch_c$$

$$2(\text{area } \triangle ABC) = ah_a = bh_b = ch_c$$

From Theorem 7.9, we have area $\triangle ABC = sr$. Therefore, by substitution:

$$2sr = ah_a = bh_b = ch_c$$

$$\frac{2s}{\frac{1}{r}} = \frac{a}{\frac{1}{h_a}} = \frac{b}{\frac{1}{h_b}} = \frac{c}{\frac{1}{h_c}}$$

$$\frac{2s}{\frac{1}{r}} = \frac{a + b + c}{\frac{1}{h_a} + \frac{1}{h_b} + \frac{1}{h_c}}$$

$$\frac{1}{\frac{1}{r}} = \frac{1}{\frac{1}{h_a} + \frac{1}{h_b} + \frac{1}{h_c}}$$

Therefore:

$$\frac{1}{r} = \frac{1}{h_a} + \frac{1}{h_b} + \frac{1}{h_c} \; \bullet$$

This result and Theorem 7.12 give us our next theorem.

THEOREM 7.14 The sum of the reciprocals of the lengths of the altitudes of a triangle equals the sum of the reciprocals of the lengths of the exradii.

Theorem 7.14 is stated symbolically as:

$$\frac{1}{h_a} + \frac{1}{h_b} + \frac{1}{h_c} = \frac{1}{r_1} + \frac{1}{r_2} + \frac{1}{r_3}$$

To conclude our study of equiradii, we will establish a relationship between the equiradii and the circumradius of a circle. This relationship is stated as Theorem 7.15.

THEOREM 7.15   The sum of the lengths of the exradii equals the sum of the length of the inradius and four times the length of the circumradius.

Proof   The diameter $\overline{YZ}$ of the circumcircle O of $\triangle ABC$ must contain X, the midpoint of both $\overline{BC}$ and $\overline{MM_1}$ (Theorem 7.5). Therefore $\overline{YZ} \perp \overline{BC}$ (and $\overline{YZ} \perp \overline{M_2M_3}$) (see Figure 7-4).

Because $\overline{YX}$ is a median of the trapezoid $M_3I_3I_2M_2$:

$$YX = \frac{1}{2}(I_2M_2 + I_3M_3) = \frac{1}{2}(r_2 + r_3)$$

Because the length of the segment joining the midpoints of the diagonals of a trapezoid equals one-half the difference of the lengths of the bases of the trapezoid, we have for trapezoid MIM_1I_1:

$$XZ = \frac{1}{2}(M_1I_1 - MI) = \frac{1}{2}(r_1 - r)$$

For circumradius R, we have:

$$2R = YX + XZ$$
$$2R = \frac{1}{2}(r_2 + r_3) + \frac{1}{2}(r_1 - r)$$
$$4R = r_1 + r_2 + r_3 - r$$

INTERACTIVE 7-4

Drag points *A*, *B*, and *C* to change the shape of the triangle and see that the equation is always true.

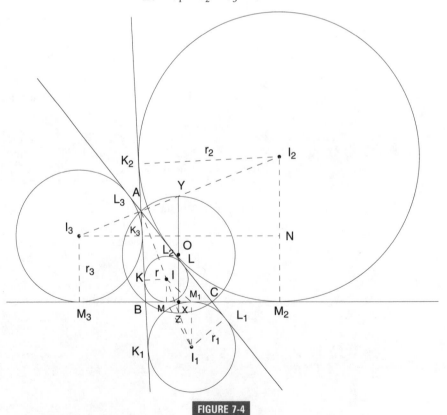

FIGURE 7-4

Therefore:

$$4R + r = r_1 + r_2 + r_3 \bullet$$

We now turn our attention to the centers of the equicircles and their distances from the circumcircle. The first theorem we state has been attributed to Leonhard Euler (1707–1783).

THEOREM 7.16   The distance, d, between the incenter and the circumcenter of a triangle is found by $d^2 = R(R - 2r)$.

Proof   Because Z is the midpoint of $\overarc{BC}$, $\overrightarrow{AI}$ must contain point Z (see Figure 7-5). Draw $\overleftrightarrow{IO}$ to intersect circle O at points D and E. Let $IO = d$. Then:

$$(AI)(IZ) = (DI)(IE) = (R - d)(R + d)$$

Consider quadrilateral $BICI_1$. Because $\overline{BI} \perp \overline{BI_1}$ (bisectors of adjacent supplementary angles) and $\overline{CI} \perp \overline{CI_1}$, quadrilateral $BICI_1$ is cyclic.

INTERACTIVE 7-5

Drag points *A, B,* and *C* to change the shape of the triangle and see that the equation is always true.

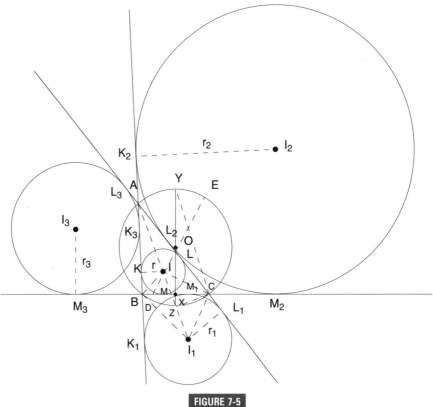

FIGURE 7-5

The center of this quadrilateral is determined by the intersection of the pependicular bisector of $\overline{BC}$ (i.e., $\overline{OXZ}$) and the diameter $\overline{II_1}$. This point of intersection is Z. Therefore $IZ = CZ$. By substitution:

$$(AI)(CZ) = R^2 - d^2 \qquad (I)$$

Because $m\angle CYZ = \frac{1}{2} m\widehat{CZ} = m\angle CAZ$ and $m\angle BAZ = m\angle CAZ$, we have:

$$m\angle CYZ = m\angle BAZ$$

Therefore:

$$\text{right triangle } YZC \sim \text{right triangle } AIK \Rightarrow \frac{AI}{YZ} = \frac{IK}{CZ} \quad \text{or} \quad (AI)(CZ) = (IK)(YZ)$$

Thus:

$$(AI)(CZ) = (r)(2R) \qquad (II)$$

From (I) and (II):

$$R^2 - d^2 = 2Rr \quad \text{or} \quad d^2 = R(R - 2r)$$

This is our desired result. ●

To complete our discussion of distances between centers, we next consider the distances between the circumcenter and each of the three excenters. You will notice that this relationship is a clear analogue of the foregoing one (see Figure 7-5).

THEOREM 7.17   The distances between the circumcenter and the three excenters of a triangle are given by $(OI_1)^2 = R(R + 2r_1)$, $(OI_2)^2 = R(R + 2r_2)$, and $(OI_3)^2 = R(R + 2r_3)$.

The proof of this theorem is similar to that of Theorem 7.16 and is left as an exercise.

This chapter should have provided you with a greater understanding of the many interrelationships involving equicircles and the circumcircle, but it by no means exhausts this topic. The exercises should serve as a springboard for further investigation.

─────── EXERCISES ───────

1. Prove that if the length of the inradius of a triangle is half the length of the circumradius, then the triangle is equilateral.

2. Prove each of the following (refer to Figure 7-1):

$$\frac{1}{r_1} = \frac{1}{h_b} + \frac{1}{h_c} - \frac{1}{h_a}$$

$$\frac{1}{r_2} = \frac{1}{h_a} + \frac{1}{h_c} - \frac{1}{h_b}$$

$$\frac{1}{r_3} = \frac{1}{h_a} + \frac{1}{h_b} - \frac{1}{h_c}$$

3. Prove that the area of a right triangle equals the product of the lengths of the two segments into which the incircle divides the hypotenuse.

4. Prove that $Rr = \dfrac{abc}{4s}$.

5. Prove that $R = \dfrac{abc}{4(\text{area } \triangle ABC)}$.

6. Prove that the ratio of the area of a triangle to the area of the triangle determined by the points of tangency of the incircle equals the ratio of twice the length of the circumradius to the length of the inradius of the given triangle.

7. Prove that $r_1 = \sqrt{\dfrac{s(s-b)(s-c)}{(s-a)}}$.

8. Prove that the sum of the distances of the circumcenter from the sides of a triangle equals the sum of the length of the circumradius and the length of the inradius of the triangle.

9. Prove Theorem 7.17.

10. Prove that if a line containing a vertex of a triangle intersects two of the equicircles, then the product of the distances of two of the points of intersection from that vertex equals the product of the distances of the other two points of intersection from that vertex.

11. Prove that the lines tangent to the incircle of a triangle and parallel to the sides of the triangle cut off three small triangles the sum of whose perimeters is equal to the perimeter of the original triangle.

12. Prove that the sum of the lengths of the legs of a right triangle minus the length of the hypotenuse equals the diameter of the incircle.

13. Prove each of the following:

$$h_a = \frac{2rr_1}{r_1 - r}$$

$$h_b = \frac{2rr_2}{r_2 - r}$$

$$h_c = \frac{2rr_3}{r_3 - r}$$

14. Prove each of the following:

$$h_a = \frac{2r_2 r_3}{r_2 + r_3}$$

$$h_b = \frac{2r_1 r_3}{r_1 + r_3}$$

$$h_c = \frac{2r_1 r_2}{r_1 + r_2}$$

15. Prove that $(OI)^2 + (OI_1)^2 + (OI_2)^2 + (OI_3)^2 = 12R^2$ (see Figure 7-5).

16. Prove that $(II_1)^2 + (II_2)^2 + (II_3)^2 = 8R(2R - r)$ (see Figure 7-5).

17. Prove each of the following, where r_a, r_b, and r_c are the radii of the excircles tangent to sides a, b, and c, respectively:

$$r_a = \frac{rs}{s - a} = \sqrt{\frac{s(s - b)(s - c)}{s - a}}$$

$$r_b = \frac{rs}{s - b} = \sqrt{\frac{s(s - a)(s - c)}{s - b}}$$

$$r_c = \frac{rs}{s - c} = \sqrt{\frac{s(s - a)(s - b)}{s - c}}$$

THE

NINE-POINT CIRCLE

INTRODUCTION TO THE NINE-POINT CIRCLE

Perhaps one of the true joys in geometry is to observe how one configuration can produce a seemingly endless array of properties and relationships. One such situation begins with nine specific points of a triangle. These points, for any given triangle, are:

- the midpoints of the sides
- the feet of the altitudes
- the midpoints of the segments from the orthocenter to the vertices

These points have the surprising relationship of all being on the same circle. This circle is called the *nine-point circle* of the triangle (see Figure 8-1).

After proving that these nine points are, in fact, concyclic, we will investigate many properties of this famous circle. In doing so, we will have to digress a bit to develop some rather interesting properties of the altitudes of a triangle.

In 1765, Leonhard Euler showed that six of these points, the midpoints of the sides and the feet of the altitudes, determine a unique circle. Yet not until 1820, when a paper published by Charles-Julian Brianchon and Jean-Victor

INTERACTIVE 8-1

Drag points *A, B,* and *C* to change the shape of the triangle. Notice that the nine points are always on the circle.

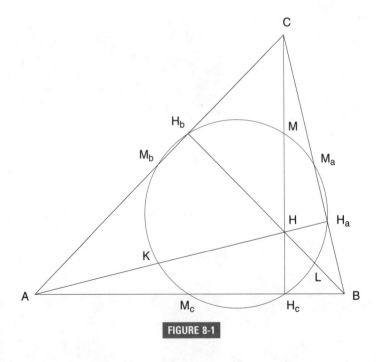

FIGURE 8-1

Poncelet appeared,* were the remaining three points (the midpoints of the segments from the orthocenter to the vertices) found to be on this circle. Their paper contains the first complete proof of the theorem and uses the name "the nine-point circle" for the first time.

Much of the fame of the gifted German mathematician Karl Wilhelm Feuerbach (1800–1834) rests on a paper he published in 1822,[†] in which he stated that "the circle which passes through the feet of the altitudes of a triangle touches all four of the circles which are tangent to the three sides of the triangle; it is internally tangent to the inscribed circle and externally tangent to each of the circles which touch the sides of the triangle externally." (This is stated in this chapter as Theorem 8.18.) As a result of his work, the theorem is referred to as the Feuerbach theorem; the nine-point circle theorem is also sometimes called the Feuerbach theorem.

THEOREM 8.1 In any triangle, the midpoints of the sides, the feet of the altitudes, and the midpoints of the segments from the orthocenter to the vertices lie on a circle.

Proof In order to simplify the discussion of this proof, we will consider each part with a separate diagram. Bear in mind that each of Figures 8-2 to 8-5 is merely an extraction from Figure 8-6, which is the complete diagram.

In Figure 8-2, points A', B', and C' are the midpoints of the three sides of $\triangle ABC$ opposite each respective vertex. $\overline{CF}$ is an altitude of $\triangle ABC$. Because $\overline{A'B'}$ is a midline of $\triangle ABC$, $\overline{A'B'} \parallel \overline{AB}$. Therefore quadrilateral $A'B'C'F$ is a trapezoid. $\overline{B'C'}$ is also a midline of $\triangle ABC$, so $B'C' = \frac{1}{2}BC$. Because $\overline{A'F}$ is the median to the hypotenuse of right triangle BCF, $A'F = \frac{1}{2}BC$. Therefore $B'C' = A'F$ and trapezoid $A'B'C'F$ is isosceles.

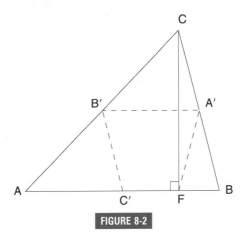

FIGURE 8-2

---

* *Recherches sur la determination d'une hyperbole équilatèau moyen de quartes conditions données* (Paris, 1820).
† *Eigenschaften einiger merkwürdigen Punkte des geradlinigen Dreiecks und mehrere durch sie bestimten Linien und Figuren. Eine analytische-trigonometrische Abhandlung* (Nürnberg, 1822).

Recall that when the opposite angles of a quadrilateral are supplementary, as in the case of an isosceles trapezoid, the quadrilateral is cyclic. Therefore quadrilateral $A'B'C'F$ is cyclic. So far we have four of the nine points on one circle.

To avoid confusion, we redraw $\triangle ABC$ and include altitude $\overline{AD}$ (Figure 8-3). Using the same argument as before, we find that quadrilateral $A'B'C'D$ is an isosceles trapezoid and therefore is cyclic. We now have five of the nine points on one circle (i.e., points A', B', C', F, and D).

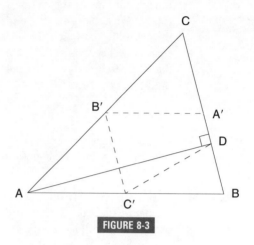

FIGURE 8-3

By repeating the same argument for altitude $\overline{BE}$ (see Figure 8-4), we can state that points D, F, and E lie on the same circle as points A', B', and C'. These six points are as far as Euler got with the configuration.

With point H as the orthocenter (the point of intersection of the altitudes), we let M be the midpoint of $\overline{CH}$ (see Figure 8-4). Therefore $\overline{B'M}$, a midline of $\triangle ACH$, is parallel to $\overline{AH}$, or altitude $\overline{AD}$ of $\triangle ABC$. Because $\overline{B'C'}$ is a midline of $\triangle ABC$, $\overline{B'C'} \parallel \overline{BC}$. Therefore, because $\angle ADC$ is a right angle, $\angle MB'C'$ is also a right angle. Thus quadrilateral $MB'C'F$ is cyclic (opposite angles are supplementary). This places point M on the circle determined by points B', C', and F. We now have a seven-point circle.

We repeat this procedure with point L, the midpoint of $\overline{BH}$ (see Figure 8-5). As before, $\angle B'A'L$ is a right angle, as is $\angle B'EL$. Therefore points B', E, A', and L

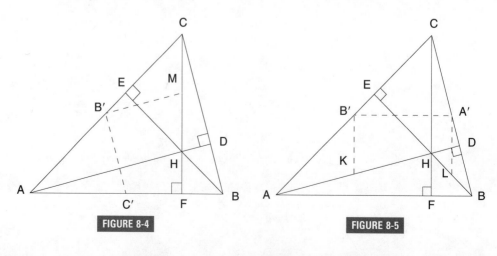

FIGURE 8-4

FIGURE 8-5

cyclic - angles are supplementary

are concyclic (opposite angles are supplementary). We now have point L as an additional point on our circle, making it an eight-point circle.

To locate our final point on the circle, consider point K, the midpoint of $\overline{AH}$. As we did earlier, we find $\angle A'B'K$ to be a right angle, as is $\angle A'DK$. Therefore quadrilateral $A'DKB'$ is cyclic and point K is on the same circle as points B', A', and D. We have proved that *nine specific points lie on this circle* (see Figure 8-6). ●

INTERACTIVE 8-6

Drag points *A, B,* and *C* to change the shape of the triangle. Notice that the center of the circle is always the midpoint of the segment.

K M L
Euler points

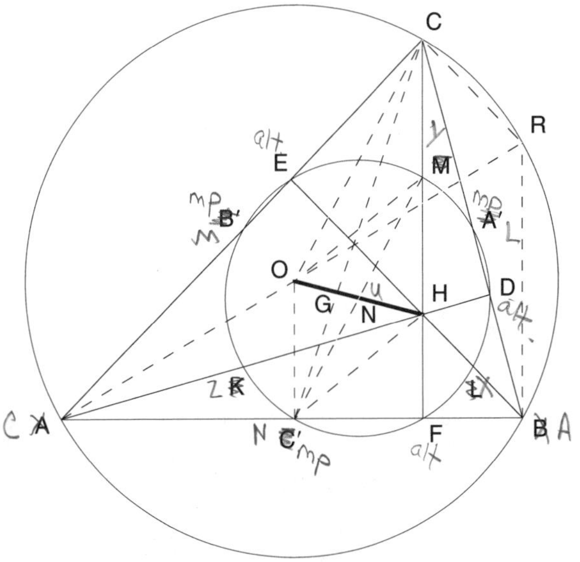

FIGURE 8-6

We are now ready to establish some basic properties of the nine-point circle.

THEOREM 8.2   The center of the nine-point circle of a triangle is the midpoint of the segment from the orthocenter to the center of the circumcircle.

P roof   Because $\overline{MC'}$ subtends a right angle at point F, it must be the diameter of the nine-point circle. Therefore the midpoint, N, of $\overline{MC'}$ is the center of the nine-point circle (see Figure 8-6).

Draw $\overrightarrow{AO}$ to intersect circumcircle O at point R. Then draw $\overline{CR}$ and $\overline{BR}$. $\overline{OC'}$ is a midline of $\triangle ARB$; therefore $\overline{OC'} \parallel \overline{RB}$. Because $\angle ABR$ is inscribed in a semicircle, it is a right angle. Both $\overline{RB}$ and $\overline{CF}$ are perpendicular to $\overline{AB}$, so $\overline{RB} \parallel \overline{CF}$. Similarly, $\overline{BE} \parallel \overline{CR}$. We therefore have parallelogram $CRBH$, so $RB = CH$.

$$OC' = \frac{1}{2} RB \ (\overline{OC'} \text{ is a midline of } \triangle ARB)$$

Therefore $OC' = \frac{1}{2}(CH) = MH$, and $OC'HM$ is a parallelogram (one pair of sides both congruent and parallel).

Because the diagonals of a parallelogram bisect each other, the midpoint, N, of $\overline{MC'}$ is also the midpoint of $\overline{OH}$. ●

▮ THEOREM 8.3 The length of the radius of the nine-point circle of a triangle is one-half the length of the radius of the circumcircle.

℗roof In Figure 8-6, we notice that $\overline{MN}$ is a midline of $\triangle OHC$. Therefore:

$$MN = \frac{1}{2}(OC)$$

This proves Theorem 8.3. ●

In a paper published in 1765, Leonhard Euler proved that the centroid, G, of a triangle trisects the segment $\overline{OH}$; that is, $OG = \frac{1}{3}(OH)$. This line, $\overrightarrow{OH}$, is known as the *Euler line* of a triangle.

▮ THEOREM 8.4 The centroid of a triangle trisects the segment from the orthocenter to the circumcenter.

℗roof We have already proved that $\overline{OC'} \parallel \overline{CH}$ (see Figure 8-6) and that $OC' = \frac{1}{2}(CH)$. Therefore $\triangle OGC' \sim \triangle HGC$ (AA) with a ratio of similitude of $\frac{1}{2}$. Therefore $OG = \frac{1}{2}(GH)$, which may be stated as $OG = \frac{1}{3}(OH)$.

It remains for us to prove that point G is the centroid of $\triangle ABC$. From the triangles we just proved similar, we have:

$$C'G = \frac{1}{2}(GC) = \frac{1}{3}(C'C)$$

Because $\overline{C'C}$ is a median, point G must be the centroid because it appropriately trisects the median. ●

It is interesting to note that:

$$\frac{HN}{NG} = \frac{3}{1} = \frac{HO}{OG}$$

Thus $\overline{HG}$ is divided internally by N and externally by O in the same ratio. This is known as a *harmonic division*.

▮ THEOREM 8.5 All triangles inscribed in a given circle and having a common orthocenter also have the same nine-point circle.

℗roof Because all triangles inscribed in a given circle and having a common orthocenter also must have the same Euler line, the center of the nine-point circle for all these triangles is fixed at the midpoint of the Euler line (Theorem 8.2). Because the length of the radius of the nine-point circle for each of these triangles is half the length of the circumradius (Theorem 8.3), they all have their nine-point circle with the same radius as well as a fixed center. Thus they all must have the same nine-point circle. ●

We now digress a bit to study a few properties of the altitudes of a triangle that will enable us to prove some other interesting relationships involving the nine-point circle.

ALTITUDES

INTERACTIVE 8-7

Drag points *A*, *B*, and *C* to change the shape of the triangle. Notice that the three triangles formed remain similar to the original triangle.

In Chapter 2, we used Ceva's theorem to prove that the altitudes of a triangle are concurrent at a point known as the *orthocenter* of the triangle (see Application 2, page 32). Recall that a triangle determined by the feet of the perpendiculars from a given point to the sides of the triangle is called a *pedal triangle*. A special kind of pedal triangle is one in which the given point from which perpendiculars are drawn to the sides is the orthocenter. This special pedal triangle is known as an *orthic triangle* (i.e., a triangle determined by the feet of the altitudes). In Figure 8-7, △*DEF* is an orthic triangle.

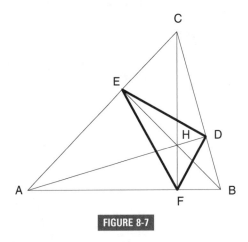

FIGURE 8-7

▌THEOREM 8.6   The orthic triangle partitions the original triangle into three additional triangles, each of which is similar to the original triangle.

In Figure 8-7, we have orthic triangle *DEF*. According to Theorem 8.6, △*ABC* is similar to each of △*DEC*, △*AEF*, and △*DBF*.

Ⓟroof   We will prove that △*ABC* ~ △*DEC* (Figure 8-7). You need only repeat this proof for △*AEF* and △*DBF* to prove Theorem 8.6 completely.
Quadrilateral *AEDB* is cyclic because ∠*AEB* and ∠*ADB* are right angles. Therefore ∠*EAB* is supplementary to ∠*EDB* (opposite angles of a cyclic quadrilateral). However, ∠*EDC* is also supplementary to ∠*EDB*. Therefore ∠*EAB* ≅ ∠*EDC*. Thus △*ABC* ~ △*DEC* (because both triangles also share ∠*ECD*).

Simply repeat this procedure to prove $\triangle ABC \sim \triangle AEF$ (use cyclic quadrilateral *ECBF*) and $\triangle ABC \sim \triangle DBF$ (use cyclic quadrilateral *AFDC*). ●

The property of an orthic triangle established as Theorem 8.6 leads us to an even more intriguing fact about orthic triangles. Consider a triangle each of whose vertices lies on the sides of a second triangle. Such a triangle is called an *inscribed triangle* of the second triangle. Now consider the possible inscribed triangles that a given acute triangle can have. Of these, the one with the shortest perimeter is the orthic triangle. We state this as Theorem 8.7.

▌THEOREM 8.7   For an acute triangle, the inscribed triangle with the minimum perimeter is the orthic triangle (see Figure 8-8).

INTERACTIVE 8-8

Drag points *A*, *B*, and *C* to change the shape of the triangle. Drag points *A′*, *B′*, and *C′* and see that the orthic triangle always is the minimum perimeter inscribed triangle.

The proof of this theorem relies heavily on a theorem from high school geometry: *The shortest distance from a given point to a given line to another given point on the same side of the line is the path that forms congruent angles with the given line.* For example, consider the points *A* and *B* on the same side of line ℓ (see Figure 8-9). Let *A′* be the reflection of point *A* in line ℓ so that $\overline{AA'} \perp \ell$ and $AR = A'R$. The intersection of $\overline{A'B}$ and ℓ is point *P*. We can easily show that $AP + PB$ is the shortest distance from point *A* to ℓ to point *B*.

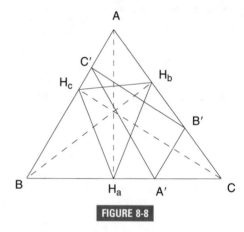

FIGURE 8-8

Because $\triangle ARP \cong \triangle A'RP$, $\angle APR \cong \angle A'PR$. Then $\angle A'PR \cong \angle BPS$. Thus $\angle APR \cong \angle BPS$, which is an important property of this minimum perimeter.

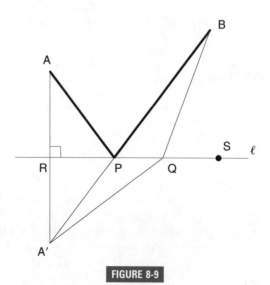

FIGURE 8-9

To prove that $AP + PB$ is the shortest required distance, select point Q on line ℓ distinct from point P. Now $A'B < A'Q + QB$, or $A'P + PB < A'Q + QB$, which verifies the selection of point P as the point determining the minimum distance, as required. We are now ready to prove Theorem 8.7.

Proof   In the proof of Theorem 8.6, we established that the orthic triangle determined three triangles similar to the original triangle. From this we can easily show that $\angle AEF \cong \angle CED$ (see Figure 8-7), $\angle CDE \cong \angle BDF$, and $\angle AFE \cong \angle BFD$.

Therefore the shortest path from point E to $\overleftrightarrow{AB}$ to point D is $EF + FD$. Similarly, the shortest path from point E to $\overleftrightarrow{CB}$ to point F is $ED + DF$, and the shortest path from point D to $\overleftrightarrow{AC}$ to point F is $DE + EF$. This implies that $\triangle EDF$ is the inscribed triangle of acute triangle ABC with the minimum perimeter.

Were we to compare the perimeter of orthic triangle DEF to that of any other inscribed triangle of $\triangle ABC$, we could easily show that $\triangle DEF$ has a smaller perimeter. ●

From the congruent angles we established earlier, another interesting property of an orthic triangle evolves. Note that because $\angle AFE \cong \angle BFD$ and because $\angle EFC$ is complementary to $\angle AFE$ and $\angle DFC$ is complementary to $\angle DFB$, we have $\angle EFC \cong \angle DFC$. The general case is stated as Theorem 8.8.

THEOREM 8.8   The altitudes of an acute triangle bisect the angles of the orthic triangle.

Refer to Application 5 on page 35. It is interesting to realize how that proof can be used in this more special situation. In effect, we now have the orthocenter of $\triangle ABC$ as the incenter of the orthic triangle DEF (see Figure 8-7).

Before expanding the diagram of Figure 8-7, we consider the following simple theorem.

THEOREM 8.9   The orthocenter of a triangle partitions each altitude into two segments, with the product of the lengths of each pair of segments equal to the product of the lengths of each of the other two pairs.

Proof   Because $\triangle CDH \sim \triangle AFH$ (see Figure 8-7), $\dfrac{CH}{AH} = \dfrac{HD}{HF}$. This can be rewritten as:

$$(CH)(HF) = (AH)(HD)$$

The proof is completed by using another pair of similar right triangles in the same manner. ●

Consider the circumcircle of $\triangle ABC$ (see Figure 8-10). Let $\overrightarrow{CF}$ intersect the circumcircle O at point S. We notice that $\overline{AB}$ bisects $\overline{HS}$. This is generally stated as in our next theorem.

INTERACTIVE 8-10

Drag points *A*, *B*, and *C* to change the shape of the triangle. Notice that the side of the triangle always bisects the segment.

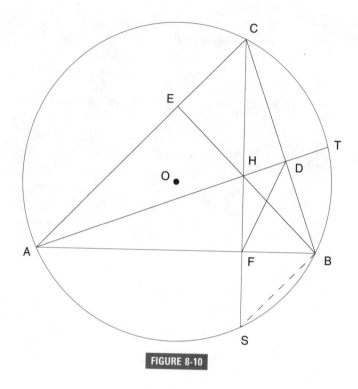

FIGURE 8-10

THEOREM 8.10   The segment from the orthocenter of a triangle to the intersection of the altitude (extended through the foot of the altitude) with the circumcircle of the triangle is bisected by a side of the triangle.

Proof   Because both angles are inscribed in the same circle and intercept the same arc, $\overset{\frown}{BC}$, $\angle CSB \cong \angle CAB$ (see Figure 8-10).

In $\triangle ACF$, $\angle ACF$ is complementary to $\angle CAF$. In $\triangle CEH$, $\angle ECH$ is complementary to $\angle CHE$. But $\angle BHF \cong \angle CHE$. Therefore $\angle CAF \cong \angle BHF$. Because both $\angle CSB$ and $\angle BHF$ are congruent to $\angle CAB$, they are congruent to each other. Thus $\triangle HBS$ is isosceles.

We can now prove $\triangle HFB \cong \triangle SFB$. It then follows that $HF = SF$, which proves the theorem for one altitude. A simple repetition of the proof may be used to verify this theorem for the other altitudes. ●

Our next theorem shows that vertex *B* is the midpoint of arc *TS* (see Figure 8-10).

■ **THEOREM 8.11**   A vertex of a triangle is the midpoint of the arc of the circumcircle determined by the intersections of two altitudes (extended through their feet) with the circumcircle (see Figure 8-11).

℗roof   Quadrilateral *AFDC* (Figure 8-10) is cyclic because ∠*AFC* and ∠*ADC* are congruent (right angles). In this cyclic quadrilateral, ∠*FAD* ≅ ∠*DCF* (both intercept $\overset{\frown}{DF}$). It then follows that $\overset{\frown}{SB} \cong \overset{\frown}{TB}$. (Congruent inscribed angles in the same circle have congruent intercepted arcs.) Remember that what holds true for one pair of altitudes can also be shown to hold true for other pairs of altitudes. ●

This configuration leads us to another pair of similar triangles.

■ **THEOREM 8.12**   The triangle formed by the intersections of the altitude extensions (through the feet of the altitudes) with the circumcircle is similar to the orthic triangle, with corresponding sides parallel.

INTERACTIVE 8-11

Drag points *A, B,* and *C* to change the shape of the triangle. Notice that the triangle formed is always similar to the orthic triangle.

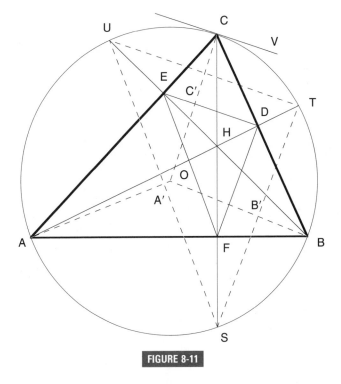

FIGURE 8-11

℗roof   In Theorem 8.10, we established that *HF* = *SF* and *HD* = *TD* (see Figure 8-11). Therefore, in △*HST*, $\overline{DF}$ is a midline and is parallel to $\overline{ST}$. The same argument is used to prove $\overline{EF} \parallel \overline{US}$ and $\overline{DE} \parallel \overline{TU}$. It then follows that △*DEF* ~ △*TUS*. ●

Theorem 8.12 leads to what will turn out to be a more useful relationship in our continued study of the nine-point circle.

THEOREM 8.13 The circumradii of a triangle that contain the vertices of the triangle are perpendicular to the corresponding sides of the orthic triangle.

Proof   We will prove this theorem for one of the radii in question and leave the proofs for the other two radii to the reader.

In Figure 8-11, $\overline{OC}$ is a radius of the circumcircles of $\triangle ABC$ and $\triangle STU$. From Theorem 8.11, $\overset{\frown}{UC} \cong \overset{\frown}{TC}$. Therefore $\overline{OC}$ is the perpendicular bisector of $\overline{TU}$. Because $\overline{OC} \perp \overline{TU}$, $\overline{OC}$ must also be perpendicular to $\overline{DE}$ (because Theorem 8.12 established $\overline{DE} \parallel \overline{TU}$). ●

We now state a theorem that follows directly from the preceding theorem.

THEOREM 8.14 The tangents to the circumcircle of a triangle at the vertices of the triangle are parallel to the corresponding sides of the orthic triangle.

Proof   Again we will prove this theorem for only one of the sides of the orthic triangle. Radius $\overline{OC}$ is perpendicular to tangent $\overleftrightarrow{VC}$ (see Figure 8-11). However, $\overline{OC} \perp \overline{DE}$ (Theorem 8.13). Therefore $\overleftrightarrow{VC} \parallel \overline{DE}$. The same argument holds for the other sides of the orthic triangle. ●

THE NINE-POINT CIRCLE REVISITED

We now return to our study of the properties of the nine-point circle. The next two properties are a direct consequence of Theorems 8.13 and 8.14.

THEOREM 8.15 Tangents to the nine-point circle of a triangle at the midpoints of the sides of the triangle are parallel to the sides of the orthic triangle.

Proof   Radius $\overline{NC'}$ of the nine-point circle is perpendicular to tangent $\overleftrightarrow{C'W}$ (see Figure 8-12). By Theorem 8.13, $\overline{OC} \perp \overline{DE}$. We showed earlier that $\overline{MN}$ was a midline of $\triangle COH$, and therefore $\overline{MN} \parallel \overline{OC}$. This implies that $\overleftrightarrow{MNC'} \parallel \overline{OC}$. Thus $\overleftrightarrow{C'W} \parallel \overline{DE}$.

The proof for the remaining two sides of the orthic triangle is done in the same manner. ●

THEOREM 8.16   Tangents to the nine-point circle at the midpoints of the sides of the given tri-angle are parallel to the tangents to the circumcircle at the opposite vertices of the given triangle.

Proof   Because the tangents to the circumcircle at a vertex of the triangle and the tangents to the nine-point circle at the midpoints of the sides of the triangle are each parallel to the sides of the orthic triangle, they are parallel to each other. ●

An *orthocentric system* consists of four points, each of which is the orthocenter of the triangle formed by the remaining three points. In Figure 8-12, points *A, B, C,* and *H* form an orthocentric system:

H is the orthocenter of △*ABC*;

A is the orthocenter of △*BCH*;

B is the orthocenter of △*ACH*;

C is the orthocenter of △*ABH*.

INTERACTIVE 8-12

Drag points *A, B,* and *C* to change the shape of the triangle. Notice that the tangents are always parallel to the sides of the orthic triangle.

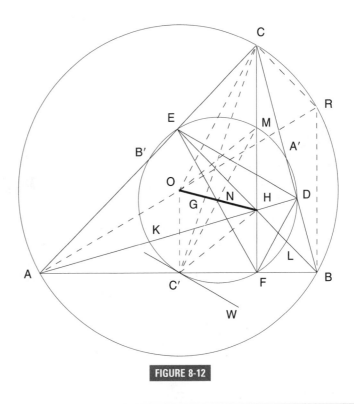

FIGURE 8-12

THEOREM 8.17   The four triangles of an orthocentric system have the same nine-point circle.

Proof   The proof of this property is left to the reader. All that is required is to check to see if, for each of the four triangles, the nine determining points all lie on the same circle *N* (Figure 8-12). ●

One of the most famous properties of the nine-point circle was first discovered (and proved) by German mathematician Karl Wilhelm Feuerbach in 1822. This property establishes a relationship between the nine-point circle and the incircle and excircles of the original triangle.

THEOREM 8.18 **(Feuerbach's theorem)** The nine-point circle of a triangle is tangent to the incircle and excircles of the triangle (see Figure 8-13).

INTERACTIVE 8-13

Drag points *A*, *B*, and *C* to change the shape of the triangle. Notice that the nine-point circle is always tangent to the two other circles.

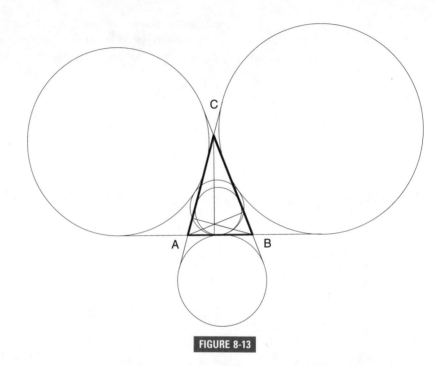

FIGURE 8-13

The proof of this property is quite complex and time-consuming. The interested reader will find four different proofs of Feuerbach's theorem in *Modern Geometry,* by Roger A. Johnson (Houghton Mifflin, 1929, pp. 200–205). The proof that Feuerbach actually used consists of computing the distances between the center of the nine-point circle and the centers of the inscribed circle (ρ), the circumscribed circle (R), and the circle inscribed in the orthic (or pedal) triangle *DEF* and showing that they equal the sum and difference of the corresponding radii:

$$(OI)^2 = R^2 - 2R\rho$$
$$(IH)^2 = 2\rho^2 - 2Rr$$
$$(OH)^2 = R^2 - 4Rr$$
$$(NI)^2 = \frac{1}{2}\left[(OI)^2 + (HI)^2\right] - (NH)^2 = \frac{1}{4}R^2 - R\rho + \rho^2 = \left(\frac{1}{2}R - \rho\right)^2$$

(Note: *I* is the center of the inscribed circle, *H* is the orthocenter, and *O* is the center of the circumscribed circle.)

—————————————— E X E R C I S E S ——————————————

1. Prove that the circumcenter of a given triangle is the orthocenter of the triangle formed by joining the midpoints of the sides of the original triangle.

2. Prove that the lengths of the altitudes of a triangle are inversely proportional to the lengths of the sides of the triangle.

3. Prove that in Figure 8-12 $\triangle A'B'C' \cong \triangle A'B'F$.

4. Prove that in Figure 8-12 $\overline{CC'}$ and $\overline{OM}$ bisect each other.

5. Why did we consider only an acute triangle when we proved that the orthic triangle had the minimum perimeter of all triangles inscribed in the original triangle (Theorem 8.7)?

6. Prove that the product of the lengths of the two segments into which the altitude partitions a side of a given triangle equals the product of the lengths of this altitude and the perpendicular segment from the orthocenter to the side.

7. Prove that the angle formed by the altitude and the circumradius containing the same vertex of a given triangle has a measure equal to the difference of the measures of the remaining two angles of the triangle.

8. Prove that the angle formed by the altitude and the circumradius containing the same vertex of a given triangle is bisected by the angle bisector of the triangle containing the vertex.

9. Prove that the circumcircle of a given triangle is congruent to the circle containing two vertices and the orthocenter of the triangle.

10. Prove that the length of the perpendicular segment from the circumcenter of a triangle to a side is equal to one-half the length of the segment from the opposite vertex to the orthocenter of the triangle.

11. Prove that the sum of the lengths of the three altitudes of a triangle is less than the perimeter of the triangle.

12. Prove that the nine-point circle of the triangle determined by any three of the incenter and excenters of the triangle is the circumcircle of the original triangle.

13. Prove that a line containing the orthocenter and the midpoint of a side of a given triangle intersects on the circumcircle the circumdiameter containing the opposite vertex.

14. Prove that the feet of the perpendiculars from the orthocenter to the exterior and interior angle bisectors of one vertex of a given triangle are collinear with the midpoint of the side opposite that vertex and the center of the nine-point circle of the triangle.

15. Prove that if a triangle has a fixed vertex and a fixed nine-point circle, then the locus of points that can be the circumcenter is a circle.

CHAPTER NINE

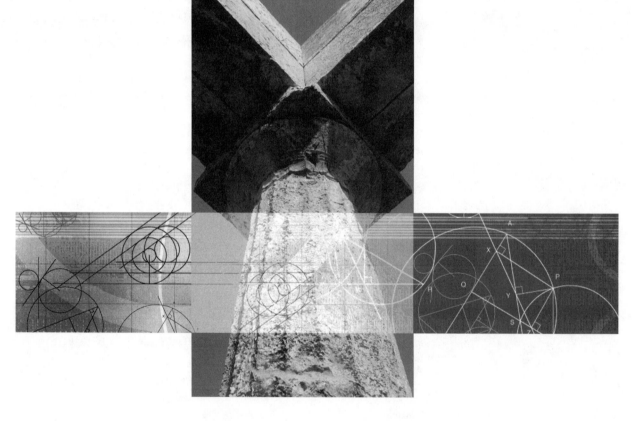

TRIANGLE CONSTRUCTIONS

INTRODUCTION

The earliest postulates on the congruence of triangles are directly related to the construction of triangles. For example, the postulate that two triangles are congruent if they agree in two sides and their included angle (SAS) is a direct consequence of the fact that, with the usual postulates on the use of traditional construction instruments (the unmarked straightedge and a pair of compasses), a *unique* triangle can be constructed if we are given two of its sides and the angle between them. This triangle is unique in the sense that, if we tried to construct another triangle from the same information (i.e., two sides and their included angle), we would inevitably (by repeating the steps in the construction) end up with a triangle that agrees with the first one in all of its parts and whose only possible difference from the first is its position in the plane. Thus we say that the given information about the triangle, SAS, *determines* the triangle. In the past, traditional constructions were made on paper. Today, however, we have in addition a computer program that produces far more accurate drawings: The Geometer's Sketchpad.

Figure 9-1 illustrates some of the details of triangles we will consider in this chapter. We list these systematically here with the general understanding that a symbol's use is clear from the context in which it is found. For instance, we may use b to denote either a side of a triangle, its name, or its measure. The ambiguity reflects our choice, not our ignorance. Our aim is clarity. The rigor and precision that support the material could certainly be supplied, but only with time and space that seem inappropriate in our discussion.

Sides: a, b, c

Angles: α, β, γ

Vertices: A, B, C

Altitudes: h_a, h_b, h_c

Feet of the altitudes: H_a, H_b, H_c

Orthocenter (point of concurrence of altitudes): H

Medians: m_a, m_b, m_c

Midpoints of sides: M_a, M_b, M_c

Centroid (point of concurrence of medians): G

Angle bisectors: t_a, t_b, t_c

Feet of angle bisectors: T_a, T_b, T_c

Incenter (center of inscribed circle, the point of concurrence of angle bisectors): I

Inradius (radius of inscribed circle): r

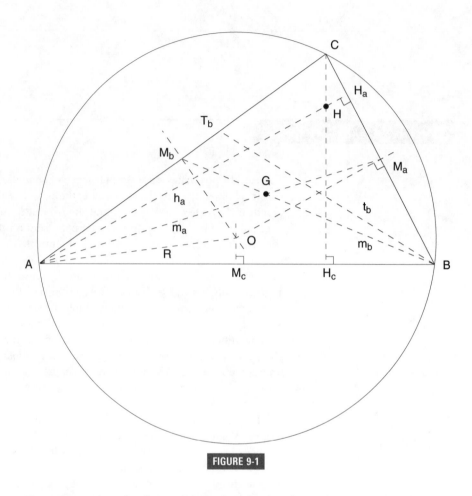

FIGURE 9-1

Circumcenter (center of circumscribed circle, the point of concurrence of the perpendicular bisectors of the sides): O

Circumradius (radius of circumscribed circle): R

Semiperimeter (half the sum of the lengths of the sides: $\frac{1}{2}(a + b + c)$): s

Note that we use a lowercase letter, in general, to represent a measure of a length and an uppercase letter to represent a point. An exception is the use of the uppercase letter R to represent the length of the circumradius in order to agree with the general use in the literature.

Much of the study of geometry is concerned with relations among the items listed. Many triangle relations (e.g., that the sum of the measures of the angles of a triangle is 180°) are already known to you. Others (e.g., Theorem 7.13, which states that the reciprocal of the length of the inradius is equal to the sum of the reciprocals of the lengths of the altitudes, or $\frac{1}{r} = \frac{1}{h_a} + \frac{1}{h_b} + \frac{1}{h_c}$) you have just learned. We will use whatever relations we may need in our constructions, with brief indications of their proofs where appropriate. You are urged to follow through on these by completing the proof if you can or looking up the proof in a more comprehensive reference. You might even find some new relations on your own.

We approach triangle constructions on two levels. On the first level, we assume that an actual triangle exists somewhere, that someone has handed us some parts of that triangle, and that our job is to *reconstruct* the original triangle. On this first level, we assume that a solution exists and that we are to find it. Our method in this case is usually to try to find a sequence of steps and appropriate relations in order to reconstruct the triangle. On the second level, we do not necessarily assume that a solution exists. We examine the given material only in light of the possibility that a solution exists, and if so, we are to determine the relation between the given information and the nature and number of possible solutions. We illustrate both approaches in solving a familiar problem: Construct a triangle given the lengths of the three sides, a, b, and c.

By the first approach, if we assume that someone actually had $\triangle ABC$ and then "took it apart" and gave us just the three lengths, a, b, and c, we can quickly reconstruct that triangle. On any line, take any point as the vertex A; then with arc (A, c) cut the line at point B, thus constructing $AB = c$ (Figure 9-2). Then draw arcs (A, b) and (B, a) to intersect at point C. Drawing segments AC and $\overline{BC}$ will give us the required triangle, $\triangle ABC$.

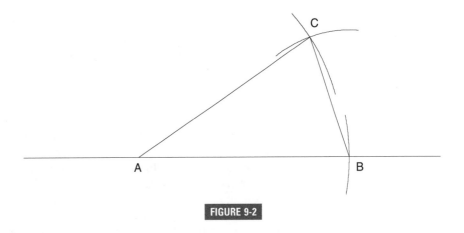

FIGURE 9-2

The second approach to the problem does not assume that a solution exists but rather "examines the gift horse in the mouth." If, for example, the given lengths are 2, 3, and 6, it should not take long to see that *no* triangle can be drawn with these measures as the lengths of its sides. Any attempt to carry through such steps will soon show the impossibility of an essential event: the intersecting of the arcs (A, b) and (B, a), without which we do not have the third vertex, C (Figure 9-3).

We are thus led to an essential requirement of any set of three lengths that we propose as the lengths of sides of a triangle: The sum of any two lengths must be greater than the third. If this condition, called the *triangle inequality*, is not satisfied, there can be *no* solution (i.e., no triangle can exist). If this condition *is* satisfied, then we *can* construct the triangle. As a matter of fact, the arcs (A, b) and (B, a) will intersect in two points, C and C', one above and the other below $\overleftrightarrow{AB}$. The two triangles we obtain in this way, $\triangle ABC$ and $\triangle ABC'$, are symmetric across $\overleftrightarrow{AB}$ and are congruent, so we have essentially one solution to this problem. (What happens if $a + b = c$?)

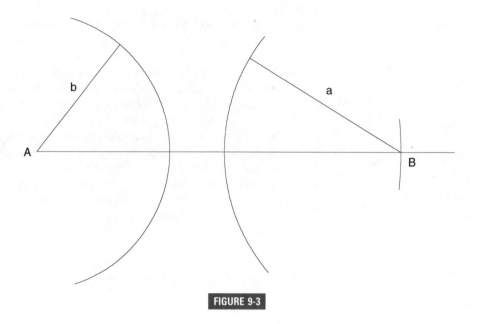

FIGURE 9-3

Discussions of the possibility and number of solutions and their relations to the given data can lead to deeper and more interesting mathematics. We will do some of this discussion in later work, and we urge you to pursue such ideas as often as possible.

Another difficulty may arise on both levels of construction. Consider, for example, the problem of constructing a triangle given the measures of its three angles $\{\alpha, \beta, \gamma\}$. If these measures came from an actual triangle, we know that the sum of their measures has to be the measure of a straight angle; that is, the measure of any one of them would be the supplement of the sum of the measures of the other two. Thus, if we are given the measures of any two of these angles, we need not be given the third because we can find it from the two that are given. A set of information for which some part need not be given because it can be found from the rest is called a *redundant set.*

In this case, we are given essentially only some information about *two* angles, say α and β, which "actually" came from $\triangle ABC$ but could as well have come from any one of infinitely many similar triangles, $\triangle AB_1C_1$, $\triangle AB_2C_2$, $\triangle AB_3C_3$, ... (Figure 9-4).

If, on the other hand, we start with *any* three given angles $\{\alpha, \beta, \gamma\}$, there can be *no* triangle with these three given angles unless they satisfy the necessary condition that the sum of their measures is the measure of a straight angle.

It should be clear that in order to construct or reconstruct any particular triangle, we must have three *independent* pieces of information about it. Any dependencies among these pieces of information may make the set redundant and therefore insufficient to determine a triangle. Note that the set $\{a, b, c\}$ is independent because a choice of a and b does not determine c. Of course, we are bound by the triangle inequality, which we can restate as $a - b < c < a + b$. (Can you show how the first kind of inequality follows from the second kind?)

The set $\{\alpha, \beta, \gamma\}$ is quite familiar as a redundant set. We now call attention to two other, less familiar redundant sets. Because a right triangle is determined

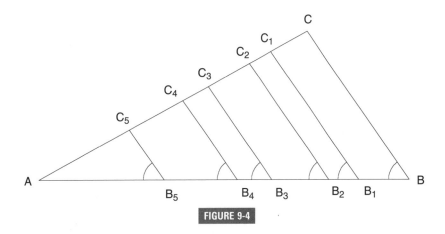

FIGURE 9-4

when we know its hypotenuse and one acute angle, it follows that the set $\{\alpha, b, h_c\}$ is a redundant set. From Figure 9-5, it should be clear that we could construct right triangle ACH_c given any two of the set $\{\alpha, b, h_c\}$. Vertex B could then be taken anywhere on $\overleftrightarrow{AH_c}$ so that we certainly have not determined any particular $\triangle ABC$. In the right triangle, we have $h_c = b \sin \alpha$, so the set $\{\alpha, b, h_c\}$ is redundant. If we are given α and b, we need not be given h_c because we can find it ourselves. We have an analogous situation if we are given $\{\alpha, \beta, h_c\}$, which is another redundant set (consider right triangle BCH_c) and does not determine any unique $\triangle ABC$.

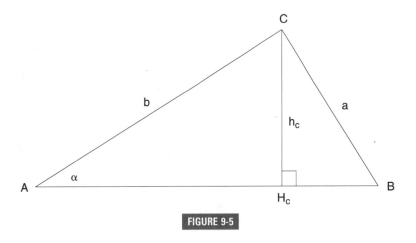

FIGURE 9-5

Another, less obvious redundant set is (a, α, R). Suppose, as in Figure 9-6, that we have drawn the circumcircle of $\triangle ABC$ and also radii $\overline{OB}$ and $\overline{OC}$ and altitude $\overline{OM_a}$ of isosceles triangle OBC. From relations involving central and inscribed angles, we can see that if α is acute, then $m\angle BOC = 2m\angle \alpha$, while if α is obtuse, then $m\angle BOC = 2m\angle(\text{supp. } \alpha')$. But from right triangle OCM_a, we have $\dfrac{a}{2} = R \sin \alpha$. Therefore, in both cases, $a = 2R \sin \alpha$. Because of this relationship, it should be clear that if we are given any two of the set (a, α, R), we can find the third ourselves. Thus this set is redundant.

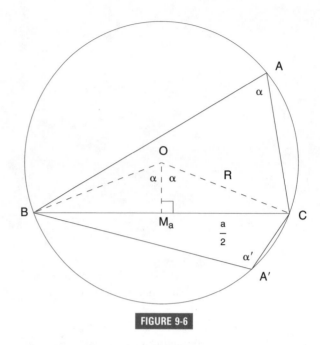

FIGURE 9-6

Following is a systematic and comprehensive listing of all 179 sets of three independent data that may determine a triangle. By changing around the letters, we could have listed each set in various other ways. Thus set 2 below, $\{a, b, \alpha\}$, which can be verbalized as "two sides and the angle opposite one of them," could have been represented by other choices of sides and angles, as long as we have two sides and the angle opposite one of them—for example, $\{a, b, \beta\}$, $\{a, c, \alpha\}$, $\{a, c, \gamma\}$, $\{b, c, \beta\}$, or $\{b, c, \gamma\}$. If you make up or come across a construction problem of this type, you can find it in this listing by writing the given information in the order that we use in each set: sides, angles, altitudes, medians, angle bisectors, circumradius, inradius, and semiperimeter.

1. $\{a, b, c\}$

2. $\{a, b, \alpha\}$

3. $\{a, b, \gamma\}$

4. $\{a, \alpha, \beta\}$

5. $\{a, b, h_a\}$

6. $\{a, b, h_c\}$

7. $\{a, \alpha, h_a\}$

8. $\{a, \alpha, h_b\}$

9. $\{a, \beta, h_a\}$

10. $\{a, \beta, h_b\}$

11. $\{\alpha, \beta, h_a\}$

12. $\{a, h_a, h_b\}$

13. $\{a, h_b, h_c\}$

14. $\{\alpha, h_a, h_b\}$

15. $\{\alpha, h_b, h_c\}$

16. $\{h_a, h_b, h_c\}$

17. $\{a, b, m_a\}$

18. $\{a, b, m_c\}$

19. $\{a, \alpha, m_a\}$

20. $\{a, \alpha, m_b\}$

21. $\{a, \beta, m_a\}$

22. $\{a, \beta, m_b\}$

23. $\{a, \beta, m_c\}$

24. $\{\alpha, \beta, m_a\}$

25. $\{a, h_a, m_a\}$

26. $\{a, h_a, m_b\}$

27. $\{a, h_b, m_a\}$

28. $\{a, h_b, m_b\}$

29. $\{a, h_b, m_c\}$

30. $\{\alpha, h_a, m_a\}$

31. $\{\alpha, h_a, m_b\}$

32. $\{\alpha, h_b, m_a\}$

33. $\{\alpha, h_b, m_b\}$

34. $\{\alpha, h_b, m_c\}$

35. $\{h_a, h_b, m_a\}$

36. $\{h_a, h_b, m_c\}$

37. $\{a, m_a, m_b\}$
38. $\{a, m_b, m_c\}$
39. $\{\alpha, m_a, m_b\}$
40. $\{\alpha, m_b, m_c\}$
41. $\{h_a, m_a, m_b\}$
42. $\{h_a, m_b, m_c\}$
43. $\{m_a, m_b, m_c\}$
44. $\{a, b, t_a\}$
45. $\{a, b, t_c\}$
46. $\{a, \alpha, t_a\}$
47. $\{a, \alpha, t_b\}$
48. $\{a, \beta, t_a\}$
49. $\{a, \beta, t_b\}$
50. $\{a, \beta, t_c\}$
51. $\{\alpha, \beta, t_a\}$
52. $\{a, h_a, t_a\}$
53. $\{a, h_a, t_b\}$
54. $\{a, h_b, t_a\}$
55. $\{a, h_b, t_b\}$
56. $\{a, h_b, t_c\}$
57. $\{\alpha, h_a, t_a\}$
58. $\{\alpha, h_a, t_b\}$
59. $\{\alpha, h_b, t_a\}$
60. $\{\alpha, h_b, t_b\}$
61. $\{\alpha, h_b, t_c\}$
62. $\{h_a, h_b, t_a\}$
63. $\{h_a, h_b, t_c\}$
64. $\{a, m_a, t_a\}$
65. $\{a, m_a, t_b\}$
66. $\{a, m_b, t_a\}$
67. $\{a, m_b, t_b\}$
68. $\{a, m_b, t_c\}$
69. $\{\alpha, m_a, t_a\}$
70. $\{\alpha, m_a, t_b\}$
71. $\{\alpha, m_b, t_a\}$
72. $\{\alpha, m_b, t_b\}$
73. $\{\alpha, m_b, t_c\}$
74. $\{h_a, m_a, t_a\}$
75. $\{h_a, m_a, t_b\}$
76. $\{h_a, m_b, t_a\}$
77. $\{h_a, m_b, t_b\}$
78. $\{h_a, m_b, t_c\}$
79. $\{m_a, m_b, t_a\}$
80. $\{m_a, m_b, t_c\}$
81. $\{a, t_a, t_b\}$
82. $\{a, t_b, t_c\}$
83. $\{\alpha, t_a, t_b\}$
84. $\{\alpha, t_b, t_c\}$
85. $\{h_a, t_a, t_b\}$
86. $\{h_a, t_b, t_c\}$

87. $\{m_a, t_a, t_b\}$
88. $\{m_a, t_b, t_c\}$
89. $\{t_a, t_b, t_c\}$
90. $\{a, b, R\}$
91. $\{a, \beta, R\}$
92. $\{\alpha, \beta, R\}$
93. $\{a, h_a, R\}$
94. $\{a, h_b, R\}$
95. $\{\alpha, h_a, R\}$
96. $\{\alpha, h_b, R\}$
97. $\{h_a, h_b, R\}$
98. $\{a, m_a, R\}$
99. $\{a, m_b, R\}$
100. $\{\alpha, m_a, R\}$
101. $\{\alpha, m_b, R\}$
102. $\{h_a, m_a, R\}$
103. $\{h_a, m_b, R\}$
104. $\{m_a, m_b, R\}$
105. $\{a, t_a, R\}$
106. $\{a, t_b, R\}$
107. $\{\alpha, t_a, R\}$
108. $\{\alpha, t_b, R\}$
109. $\{h_a, t_a, R\}$
110. $\{h_a, t_b, R\}$
111. $\{m_a, t_a, R\}$
112. $\{m_a, t_b, R\}$
113. $\{t_a, t_b, R\}$
114. $\{a, b, r\}$
115. $\{a, \alpha, r\}$
116. $\{a, \beta, r\}$
117. $\{\alpha, \beta, r\}$
118. $\{a, h_a, r\}$
119. $\{a, h_b, r\}$
120. $\{\alpha, h_a, r\}$
121. $\{\alpha, h_b, r\}$
122. $\{h_a, h_b, r\}$
123. $\{a, m_a, r\}$
124. $\{a, m_b, r\}$
125. $\{\alpha, m_a, r\}$
126. $\{\alpha, m_b, r\}$
127. $\{h_a, m_a, r\}$
128. $\{h_a, m_b, r\}$
129. $\{m_a, m_b, r\}$
130. $\{a, t_a, r\}$
131. $\{a, t_b, r\}$
132. $\{\alpha, t_a, r\}$
133. $\{\alpha, t_b, r\}$
134. $\{h_a, t_a, r\}$
135. $\{h_a, t_b, r\}$
136. $\{m_a, t_a, r\}$

137. $\{m_a, t_b, r\}$	159. $\{m_a, m_b, s\}$
138. $\{t_a, t_b, r\}$	160. $\{a, t_a, s\}$
139. $\{a, R, r\}$	161. $\{a, t_b, s\}$
140. $\{\alpha, R, r\}$	162. $\{\alpha, t_a, s\}$
141. $\{h_a, R, r\}$	163. $\{\alpha, t_b, s\}$
142. $\{m_a, R, r\}$	164. $\{h_a, t_a, s\}$
143. $\{t_a, R, r\}$	165. $\{h_a, t_b, s\}$
144. $\{a, b, s\}$	166. $\{m_a, t_a, s\}$
145. $\{a, \alpha, s\}$	167. $\{m_a, t_b, s\}$
146. $\{a, \beta, s\}$	168. $\{t_a, t_b, s\}$
147. $\{\alpha, \beta, s\}$	169. $\{a, R, s\}$
148. $\{a, h_a, s\}$	170. $\{\alpha, R, s\}$
149. $\{a, h_b, s\}$	171. $\{h_a, R, s\}$
150. $\{\alpha, h_a, s\}$	172. $\{m_a, R, s\}$
151. $\{\alpha, h_b, s\}$	173. $\{t_a, R, s\}$
152. $\{h_a, h_b, s\}$	174. $\{a, r, s\}$
153. $\{a, m_a, s\}$	175. $\{\alpha, r, s\}$
154. $\{a, m_b, s\}$	176. $\{h_a, r, s\}$
155. $\{\alpha, m_a, s\}$	177. $\{m_a, r, s\}$
156. $\{\alpha, m_b, s\}$	178. $\{t_a, r, s\}$
157. $\{h_a, m_a, s\}$	179. $\{R, r, s\}$
158. $\{h_a, m_b, s\}$	

This list can be considered as a list of 179 construction problems, which you are invited to plunge into. We will do several of these constructions in the rest of this chapter to demonstrate some useful techniques and to develop some geometric information you may not have come across before. In the first few constructions, we will discuss rather fully the possibility and number of solutions under various conditions, but in later problems, we will leave this interesting (and more difficult) work for you.

SELECTED CONSTRUCTIONS

■ CONSTRUCTION 5 $\{a, b, h_a\}$

On any line, take point H_a and construct perpendicular H_aA of length h_a (see Figure 9-7). Let arc (A, b) intersect this base line at point C, and let arc (C, a) intersect this base line at points B and B'. The two solutions are $\triangle ABC$ and $\triangle AB'C$, each of which has the given $\{a, b, h_a\}$.

Discussion   Because arc (A, b) must intersect the base line to obtain point C, a necessary condition for a solution is that $b \geq h_a$. Because this arc will intersect the line again at point C' (to the right of point H_a), we will also have another pair of

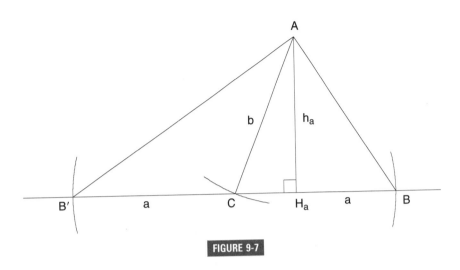

FIGURE 9-7

solutions, but these will be reflections of the solutions we already have. Another possibility would follow if we had taken the perpendicular at point H_a below as well as above the base line, but again we would have reflections of the solutions we found before. In later work, we will not discuss such reflections or symmetric solutions that contribute nothing essentially new. If $b = h_a$, then there will be just one point of contact between arc (A, b) and the base line, at the point H_a itself, which is then a point of tangency of this arc. In this case, triangles ACB and ACB' become congruent right triangles; that is, we have essentially a single solution. If $b > h_a$, we get two solutions no matter what length is chosen for a. Thus, finally, the condition $b \geq h_a$ is necessary and sufficient for any solution to this problem, with the equality leading to one solution and the inequality leading to two solutions. ●

CONSTRUCTION 7 $\{a, \alpha, h_a\}$

This problem is nicely done by the intersection of loci. On any line, construct $BC = a$ (see Figure 9-8). Then one locus for vertex A is a line parallel to $\overleftrightarrow{BC}$ at distance h_a, that is, L_1, because every point on L_1 is at distance h_a from this base line. Another locus for vertex A is the circular arc that subtends $\overline{BC}$ and in which any inscribed angle would have measure equal to α. Thus, to solve this problem, we construct L_1 and L_2 as indicated to intersect at point A and then draw $\overline{AB}$ and $\overline{AC}$ to finish the solution triangle ABC. (Another solution, congruent to this one, would come from A', the other intersection of the loci.)

Discussion We will have solutions if and only if the two loci intersect, which will occur if h_a is not "too big." Consider the "tallest" triangle, $\triangle A''BC$, and in particular right triangle $A''BM_a$, in which $m\angle BA''M_a = \frac{1}{2}\alpha$ and $BM = \frac{1}{2}a$. The original problem will have a solution if and only if $h_a \leq A''M_a$, which becomes, from relations in that right triangle, $h_a \leq \frac{1}{2}a \cot \frac{1}{2}\alpha$. With equality in this situation, we will have L_1 tangent to L_2 at point A'' and the only solution will be isosceles triangle $A''BC$;

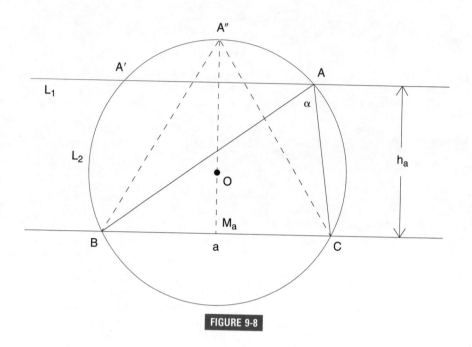

FIGURE 9-8

with inequality, we will have two congruent triangles, $\triangle ABC$ and $\triangle A'BC$, or essentially one solution. ●

CONSTRUCTION 13 $\{a, h_b, h_c\}$

On any line, construct $BC = a$ and a semicircle with $\overline{BC}$ as diameter (Figure 9-9). This semicircle is a locus for both H_b and H_c because both $\angle BH_bC$ and $\angle BH_cC$ are right angles. Now draw arc (B, h_b) to intersect this semicircle at

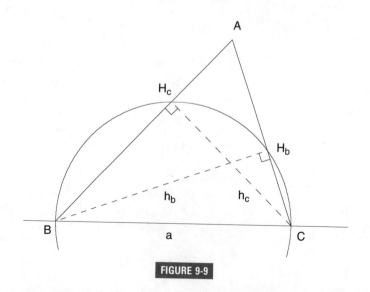

FIGURE 9-9

point H_b and arc (C, h_c) to intersect it at point H_c. Finally, $\overleftrightarrow{BH_c}$ and $\overleftrightarrow{CH_b}$ intersect at point A to give us the solution triangle ABC.

We leave the full discussion to the reader, with a hint that you should examine the relative lengths a, h_a, and h_c that determine the intersections of the various arcs that enter into the construction.

CONSTRUCTION 16 $\{h_a, h_b, h_c\}$

Solution I   Because the area of $\triangle ABC$ is $\frac{1}{2}ah_a$, which equals $\frac{1}{2}bh_b$, which equals $\frac{1}{2}ch_c$, we have $ah_a = bh_b = ch_c$, from which we could write:

$$a:\frac{1}{h_a} = b:\frac{1}{h_b} = c:\frac{1}{h_c}$$

These equations tell us that the lengths of the sides of a triangle are inversely proportional to the lengths of their corresponding altitudes. Conversely, we also have:

$$h_a:\frac{1}{a} = h_b:\frac{1}{b} = h_c:\frac{1}{c}$$

If we make a new $\triangle PQR$, with sides h_a, h_b, and h_c (Figure 9-10), the lengths of the altitudes of $\triangle PQR$, h'_a, h'_b, and h'_c, will also be inversely proportional to the lengths of the sides of the new triangle, h_a, h_b, and h_c:

$$h_a:\frac{1}{h'_a} = h_b:\frac{1}{h'_b} = h_c:\frac{1}{h'_c}$$

But we can see that the lengths of these new altitudes will be directly proportional to the lengths of the sides of the original triangle ABC:

$$a:h'_a = b:h'_b = c:h'_c$$

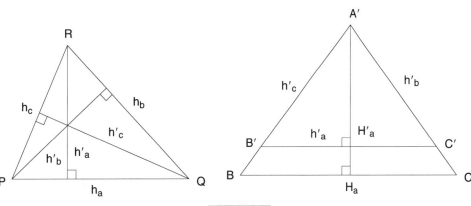

FIGURE 9-10

Therefore a new triangle whose sides have length h_a', h_b', and h_c' will be similar to the required triangle ABC.

The construction then follows these steps: (1) Construct $\triangle PQR$ whose sides are the given altitudes, h_a, h_b, and h_c; (2) find the altitudes h_a', h_b', and h_c' of $\triangle PQR$; (3) construct $\triangle A'B'C'$ whose sides are the altitudes just found ($\triangle A'B'C'$ is similar to the solution triangle ABC) (see Figure 9-10); (4) construct any altitude, say $\overline{A'H_a'}$, of $\triangle A'B'C'$, and on $\overrightarrow{A'H_a'}$ construct $\overline{A'H_a}$ congruent to the given altitude of length h_a; (5) through H_a draw a perpendicular to $A'H_a$ intersecting $\overleftrightarrow{A'B'}$ and $\overleftrightarrow{A'C'}$ at points B and C. Then $\triangle A'BC$ is the required triangle. ●

Solution II   If secants are drawn from a point E to a circle, then $EF \cdot EG = EK \cdot EL = EM \cdot EN$, and so on (Figure 9-11). These equal products can be related to the equal products $ah_a = bh_b = ch_c$ with the following construction: From any point E outside any circle, draw secants to the circle so that $EF = h_a$, $EK = h_b$, and $EM = h_c$, as shown in Figure 9-11. These secants will intersect the same circle again at points G, L, and N, respectively, with the corresponding lengths

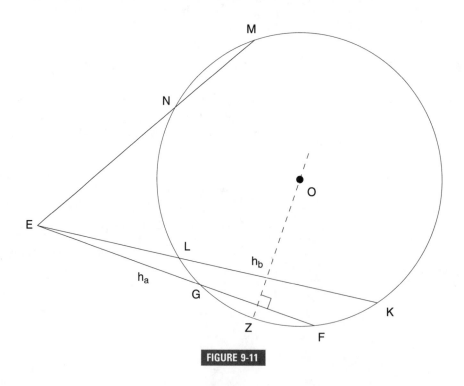

FIGURE 9-11

$EG = a'$, $EL = b'$, and $EN = c'$. By construction, $a'h_a = b'h_b = c'h_c$, and by division we obtain $a : a' = b : b' = c : c'$. Thus a triangle whose sides have lengths a', b', and c', just found, will be similar to the solution triangle ABC. We continue the construction by making $\triangle A'B'C'$ with sides congruent to $\overline{EG}$, $\overline{EL}$, and $\overline{EN}$ and then proceed as in Solution I. In both cases, $\triangle A'B'C'$ will be similar to the solution triangle ABC. ●

CONSTRUCTION 20 $\{a, \alpha, m_b\}$

On any line, construct $BC = a$ and then construct circular arc BAC to contain α. This arc is a locus for the vertex A, and its circle is the circumcircle of $\triangle ABC$. Then construct the circle with diameter $\overline{OC}$. This circle, L_1, is a locus for the midpoints of all arcs of the first circle that can be drawn from point C and is thus a locus for point M_b. Because the distance from point B to point M_b is the given median length m_b, another locus for point M_b is L_2, the circle (B, m_b) that we draw. Thus point M_b is at the intersection of these loci, as shown in Figure 9-12, and $\overleftrightarrow{CM_b}$ will meet the first circle at point A. Then $\triangle ABC$ is our solution.

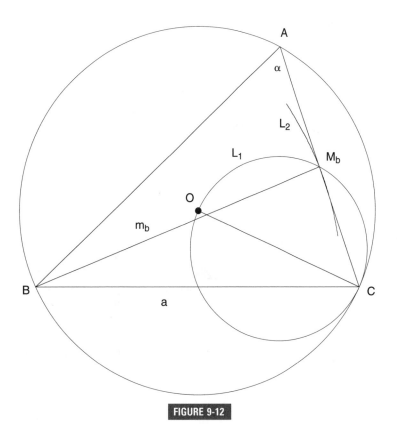

FIGURE 9-12

CONSTRUCTION 29 $\{a, h_b, m_c\}$

Suppose the triangle is constructed. If we extend $\overline{CM_c}$ its own length to point D, then quadrilateral $ACBD$ is a parallelogram because the diagonals bisect each other (Figure 9-13). The lengths a and h_b determine the right triangle BH_bC and thus the parallelogram $ACBD$. Hence the construction.

On any line, construct $CB = a$ and then construct a semicircle on diameter $\overline{BC}$. (This semicircle is a locus for point H_b). With arc (B, h_b), cut this semicircle at point H_b; then draw $\overleftrightarrow{CH_b}$ and draw $\overleftrightarrow{BD}$ through point B and parallel to $\overleftrightarrow{CH_b}$. With arc

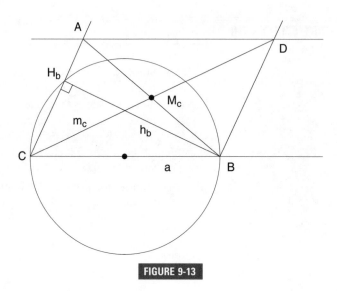

FIGURE 9-13

$(C, 2m_c)$, cut $\overleftrightarrow{BD}$ at point D, which will be a third vertex of the parallelogram. Through point D, draw a line parallel to $\overleftrightarrow{BC}$ to intersect $\overleftrightarrow{CH_b}$ at point A, the fourth vertex of the parallelogram and the third vertex of our solution triangle ABC.

CONSTRUCTION 35 $\quad \{h_a, h_b, m_a\}$

Suppose the triangle is available. Then in $\triangle AM_aH_a$, we know the hypotenuse, m_a, and leg h_a, so the triangle is determined. In right triangle AM_aK, because M_a is the midpoint of side $\overline{BC}$, its distance to side $\overline{AC}$ will be half the distance from point B to $\overline{AC}$; that is, $M_aK = \frac{1}{2}h_b$. Thus, in this right triangle AM_aK, we also know the hypotenuse length, m_a, and the length of leg $\overline{M_aK} = \frac{1}{2}h_b$. Hence the construction.

On any line, make $AM_a = m_a$ and then draw a circle with $\overline{AM_a}$ as diameter (Figure 9-14). This circle is a locus for both point H_a and point K. Cut this circle with arc (A, h_a) to locate point H_a, then with arc $(M_a, \frac{1}{2}h_b)$ to locate point K. Then $\overleftrightarrow{AK}$ and $\overleftrightarrow{H_aM_a}$ intersect at vertex C. Finally, extend $\overline{CM_a}$ its own length to point B and draw $\overline{AB}$ to complete the solution triangle ABC.

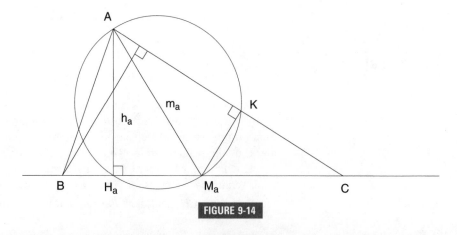

FIGURE 9-14

CONSTRUCTION 43 $\{m_a, m_b, m_c\}$

Suppose the triangle is available. Extend $\overline{GM_c}$ its own length to point D and draw $\overline{AD}$ and $\overline{DB}$ (see Figure 9-15). Then quadrilateral $ADBG$ is a parallelogram because diagonals $\overline{AB}$ and $\overline{GD}$ bisect each other. But $AG = \frac{2}{3}m_a$, $AD = BG = \frac{2}{3}m_b$, and $GD = 2(GM_c) = \frac{2}{3}m_c$. Thus $\triangle ADG$ has sides whose lengths are, respectively, two-thirds the lengths of the given medians. Hence the construction.

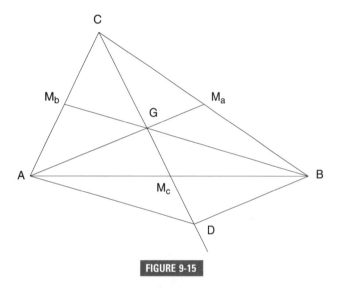

FIGURE 9-15

Construct segments whose lengths are two-thirds the lengths of the given medians, and with these segments construct $\triangle ADG$ with sides $AD = \frac{2}{3}m_b$, $DG = \frac{2}{3}m_c$, and $GA = \frac{2}{3}m_a$. Now construct median $\overline{AM_c}$ of this triangle and double its length to point B, a vertex of the required triangle. Finally, double the length of $\overline{DG}$ to point C and draw sides $\overline{AC}$ and $\overline{BC}$ of the solution triangle ABC.

CONSTRUCTION 56 $\{a, h_b, t_c\}$

Suppose the triangle is available. Then right triangle CBH_b is determined because we know its hypotenuse, a, and a leg, h_b. But in this right triangle, we have $\angle BCH_b$, which is also an angle of the solution triangle ABC. Thus we have the construction.

On any line, select the point H_b and draw a perpendicular there, making $H_bB = h_b$ (Figure 9-16). Then arc (B, a) will intersect this base line at point C, and we can draw $\overline{BC}$. Then bisect $\angle BCH_b$ and on this bisector make

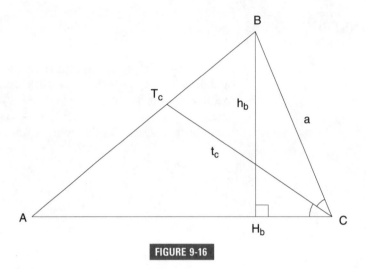

FIGURE 9-16

$CT_c = t_c$. Finally, the intersection of $\overleftrightarrow{BT_c}$ and $\overleftrightarrow{CH_b}$ is the vertex A of solution triangle ABC.

CONSTRUCTION 63 $\{h_a, h_b, t_c\}$

Suppose the triangle is available. From T_c, draw $\overline{T_c Y}$ perpendicular to $\overline{BC}$ (Figure 9-17). Now, because an angle bisector divides the opposite side into segments whose lengths are proportional to the lengths of the adjacent sides, we have $\dfrac{AT_c}{T_c B} = \dfrac{b}{a}$. Also, as previously shown, the lengths of the sides of a triangle are inversely proportional to the lengths of the corresponding altitudes. Thus we have

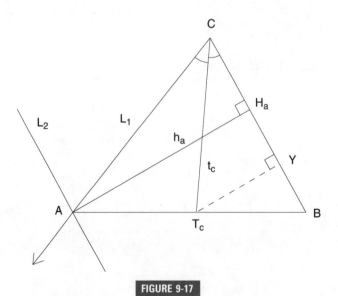

FIGURE 9-17

$\frac{b}{a} = \frac{h_a}{h_b}$. Combining these equations $\left(\text{both} = \frac{b}{a}\right)$, we have $\frac{AT_c}{T_cB} = \frac{h_a}{h_b}$. From this proportion, we obtain:

$$\frac{h_a + h_b}{h_b} = \frac{AT_c + T_cB}{T_cB} = \frac{AB}{T_cB}$$

From right triangle BAH_a and right triangle BT_cY, we also have:

$$\frac{AB}{T_cB} = \frac{AH_a}{T_cY} = \frac{h_a}{T_cY}$$

Therefore, finally:

$$\frac{h_a + h_b}{h_b} = \frac{h_a}{T_cY}$$

This shows T_cY as fourth proportional* to known quantities and therefore constructible.

The construction starts with finding the segment $\overline{T_cY}$ as fourth proportional from the available lengths, $h_a + h_b$, h_b, and h_a. Then we construct right triangle CYT_c from its known hypotenuse $CT_c = t_c$ and leg $\overline{T_cY}$. But this right triangle contains $\angle YCT_c$, which has half the measure of $\angle BCA$ of our solution triangle. If we copy $\angle YCT_c$ on the other side of $\overline{CT_c}$, then $\overrightarrow{CA}$ is a locus, L_1, for vertex A. But because this vertex is also at distance h_a from its opposite side, another locus for vertex A is L_2, the line parallel to $\overleftrightarrow{CY}$ and at distance h_a from it. These two loci intersect at vertex A, and finally the lines $\overleftrightarrow{AT_c}$ and $\overleftrightarrow{CY}$ intersect at B, the third vertex of our solution triangle ABC.

Discussion   This solution was obtained by an algebraic analysis that was far from obvious. Of course, it was necessary to know the geometric relations that led to the proportions. Our later constructions lean on even more unfamiliar geometric relations, and you are urged to deepen and extend your knowledge of geometry as you venture into the rougher waters that lie ahead. ●

CONSTRUCTION 74 $\{h_a, m_a, t_a\}$

Suppose, as usual, that the solution triangle is available. Then right triangle AH_aT_a and right triangle AH_aM_a are both determined by a known leg and hypotenuse. If we draw the circumcircle and then radius $\overline{OA}$ and radius $\overline{OM_aD}$ (Figure 9-18), we can prove a minor theorem of essential importance in this construction.

---

* The fourth proportion refers to the last part of a proportion. In the following proportion, d is called the fourth proportion: $\frac{a}{b} = \frac{c}{d}$.

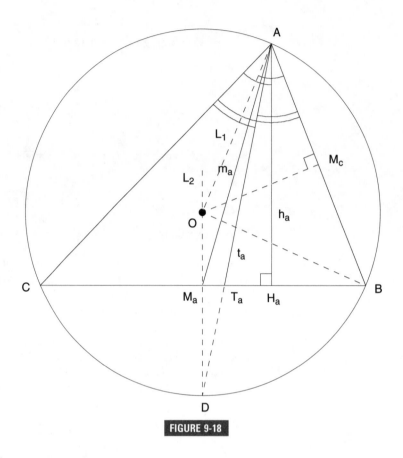

FIGURE 9-18

THEOREM An angle bisector of a triangle also bisects the angle formed by the altitude and circumradius from the same vertex.

Proof The perpendicular from circumcenter O to $\overline{AB}$ meets $\overline{AB}$ at M_c, the midpoint of $\overline{AB}$. Because central angle AOB and inscribed angle ACB intercept the same arc, we have $\gamma = \frac{1}{2}m\angle AOB = m\angle AOM_c$. Therefore, from right triangle ACH_a and right triangle OAM_c, we have $m\angle CAH_a = m\angle OAM_c$ = complement of γ. But because $\overleftrightarrow{AT_a}$ bisects $\angle BAC$, we have $m\angle BAT_a = m\angle CAT_a$. Therefore, by subtraction, we have $m\angle OAT_a = m\angle H_aAT_a$; thus $\overleftrightarrow{AT_a}$ bisects not only $\angle BAC$ but also $\angle H_aAO$. ●

The same figure leads to another useful conclusion.

THEOREM An angle bisector of a triangle meets the perpendicular bisector of its opposite side on the circumcircle of that triangle.

That is, $\overleftrightarrow{OM_a}$ and $\overleftrightarrow{AT_a}$ meet at point D, on the circumcircle (see Figure 9-18). This result follows immediately from the fact that both of these lines must bisect $\overset{\frown}{BC}$.

Our construction follows readily from the first of these two theorems. Construct right triangle AH_aT_a and right triangle AH_aM_a as usual and then double $\angle H_aAT_a$ beyond $\overleftrightarrow{AT_a}$ to obtain one locus, L_1, for the circumcenter O ($\overleftrightarrow{AT_a}$ bisects $\angle H_aAO$). Another locus for O is the perpendicular, L_2, to $\overleftrightarrow{H_aM_a}$ at point M_a. Finally, the circle (O, OA) will meet $\overleftrightarrow{H_aM_a}$ at points B and C, the other two vertices of the solution triangle ABC.

Discussion   Of course, we must have both m_a and t_a greater than h_a in order to construct our first two right triangles. ●

CONSTRUCTION 99 $\{a, m_b, R\}$

Construct isosceles triangle OBC with known sides $OB = OC = R$ and $BC = a$ (Figure 9-19). Then circle (O, R) is a locus, L_1, for the third vertex of the solution triangle ABC. But because M_b is the midpoint of $\overline{AC}$, one locus for M_b is the circle, L_1, with $\overline{OC}$ as diameter. Because M_b is at a known distance, m_b, from vertex B, another locus for M_b is L_2, the circle (B, m_b), which will intersect L_1 at M_b and M'_b. Then $\overleftrightarrow{CM_b}$ and $\overleftrightarrow{CM'_b}$ will meet the circumcircle (O, R) at the third vertices, A and A', of the solution triangles ABC and $A'BC$.

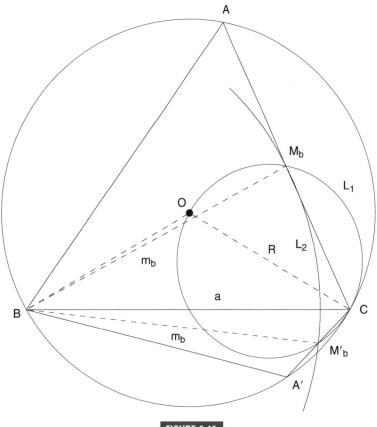

FIGURE 9-19

In this case, we have shown two distinct solutions obtained from the given data. Of course, the given lengths could have led to loci that did not intersect at all, in which case there would have been no solution.

CONSTRUCTION 102 $\{h_a, m_a, R\}$

Suppose the solution is available. Then in right triangle AH_aM_a, we know the length of hypotenuse $AM_a = m_a$ and the length of leg $AH_a = h_a$, so this triangle is determined. But the circumcenter can then be found because it is on the perpendicular bisector of $\overline{BC}$, that is, on the perpendicular to $\overleftrightarrow{M_aH_a}$ at point M_a, and it is also at known distance R from vertex A. Hence the construction.

Construct right triangle AH_aM_a with known hypotenuse length $AM_a = m_a$ and known leg length $AH_a = h_a$ (Figure 9-20). Then one locus for circumcenter O is L_1, circle (A, R). Another locus for O is L_2, the perpendicular to $\overleftrightarrow{H_aM_a}$ at point M_a. These loci intersect at the circumcenter O. Finally, with (O, R) we intersect the base line $\overleftrightarrow{H_aM_a}$ at points B and C, the other two vertices of the solution triangle ABC.

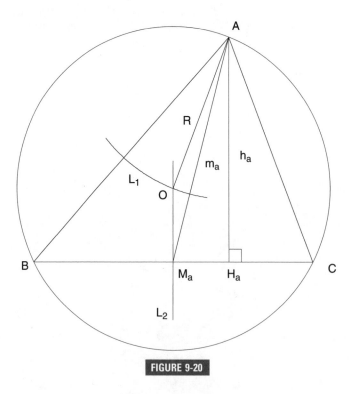

FIGURE 9-20

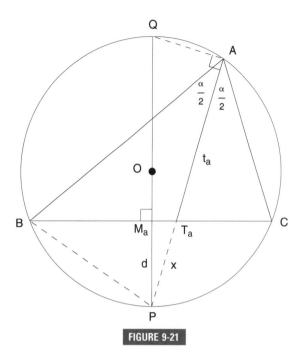

FIGURE 9-21

CONSTRUCTION 105 $\{a, t_a, R\}$

Suppose we have the solution triangle ABC and its circumcircle (O, R) available. Then the diameter perpendicular to side $\overline{BC}$ will bisect $\overline{BC}$ at point M_a and meet the circumcircle at points P and Q, as shown in Figure 9-21. The bisector of $\angle BAC$, $\overline{AT_a}$, will also meet the circumcircle at point P, the midpoint of $\overset{\frown}{BPC}$, as indicated in Construction 74. Then $\angle QAP$ is a right angle because it is inscribed in a semicircle, and the right triangles PAQ and PM_aT_a are similar because they have acute angle APQ in common. These relationships give us the proportion $\dfrac{PT_a}{PM_a} = \dfrac{PQ}{PA}$. If we name the appropriate lengths as indicated in Figure 9-21, then the proportion becomes $\dfrac{x}{d} = \dfrac{2R}{x + t_a}$.

Because $\overline{PQ}$ is the circumdiameter, its length, $2R$, is known, and of course t_a is given. The length d is readily found once we have drawn the chord of length a in the circumcircle of radius R, because d is exactly the distance from the midpoint of that chord to the midpoint of its arc. (Could you show that $d = R - \sqrt{\dfrac{R^2 - a^2}{4}}$?)

Thus the proportion leads to an equation in which x is found in terms of available lengths, d, $2R$, and t_a: $x(x + t_a) = d(2R)$. We now show a geometric solution to equations of the form $x(x + u) = vw$, where u, v, and w are known. On any line, construct $\overline{JK}$ of length v and $\overline{JL}$ of length w; then draw $\overline{LM}$ of length u in any direction. Circumscribe $\triangle KLM$ with circle π_1 and then find the distance, z, from its center to $\overline{LM}$. Now draw circle (Y, z) and draw a tangent to this circle from outside point J, to intersect circle π_1 at points D and E (see Figure 9-22).

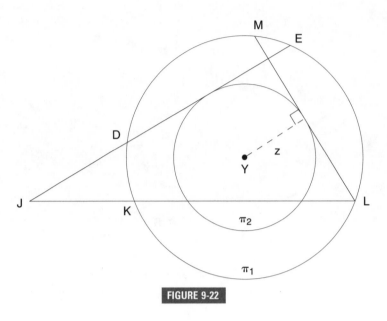

FIGURE 9-22

By the construction, chords $\overline{DE}$ and $\overline{ML}$ are equidistant from Y, the center of circle π_1, so $DE = ML$. Finally, because $JD \cdot JE = JK \cdot JL$, we have $JD \cdot (JD + DE) = JK \cdot JL$; that is, $JD \cdot (JD + u) = vw$. We have thus found $JD = x$ and $x(x + u) = vw$, as desired.

To put these steps together for our actual construction, we start by drawing circumcircle (O, R) and placing in it chord $\overline{BC}$ of given length a. Perpendicular bisector $\overline{BC}$ gives us the lengths $d = PM_a$ and $2R = PQ$ for the next construction steps, to be drawn in a separate figure.

On any line, draw $JK = PM_a = d$ and $JL = PQ = 2R$. Then from point L in any direction, draw $\overline{LM}$ of given length t_a. Circumscribe $\triangle KLM$ as described and then draw the perpendicular from the center Y to $\overline{LM}$. Then draw circle π_2, concentric with π_1 and tangent to $\overline{LM}$. Draw, as in Chapter 1, a tangent $\overleftrightarrow{JDE}$ to circle π_2, and from the theory stated earlier, we now have $JD = x$.

Back in our first figure, we draw the circle (P, x) to intersect $\overline{BC}$ at point T_a, and finally $\overleftrightarrow{PT_a}$ will meet the circumcircle at point A, the third vertex of the solution triangle ABC.

Discussion We have utilized a lot of good algebra and geometry, and we leave further details and comments to you, noting only that in the original figure the four points A, Q, M_a, and T_a all lie on a circle (with diameter $\overline{QT_a}$) because both $\angle T_aAQ$ and $\angle T_aM_aQ$ are right angles. ●

CONSTRUCTION 115 $\{a, \alpha, r\}$

This solution will also lead us through some interesting geometry that may not be familiar to you. Suppose, as usual, that the solution triangle is available. We know that the bisector of $\angle BAC$ goes through the incenter I and the midpoint P of the opposite arc, $\overset{\frown}{BPC}$, of the circumscribed circle (Figure 9-23).

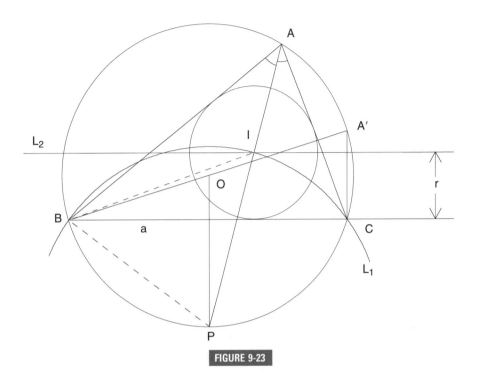

FIGURE 9-23

Consider $\triangle PBI$. Because $\angle BPI$ and $\angle BCA$ are inscribed angles both of which intersect $\overset{\frown}{AB}$, we have $m\angle BPI = m\angle BCA = \gamma$. Also, because $\overrightarrow{BI}$ and $\overrightarrow{AI}$ are both angle bisectors of the solution triangle ABC:

$$m\angle PBI = m\angle PBC + m\angle CBI = \frac{\alpha}{2} + \frac{\beta}{2}$$

Thus, because $\alpha + \beta + \gamma = 180°$, we find that $m\angle BIP = \frac{\alpha}{2} + \frac{\beta}{2}$ and therefore $\triangle PBI$ is isosceles, with $\overline{PB} \cong \overline{PI}$. But $\overline{PB}$ can be found from the given information because a and α are enough to determine the circumcircle, as indicated earlier in this chapter, and once we have placed the known chord $\overline{BC}$ in the known circumcircle (O, OB), we can easily draw radius $\overline{OP}$ as the perpendicular bisector of that chord and then draw $\overline{PB}$.

Thus we have one locus, L_1, for the incenter I: the circle (P, PB) that is available from the given information. Another locus for I follows from the fact that the incircle is tangent to each side of the solution triangle and its center, I, is thus r units distant from each side. Thus our second locus for I is the line L_2, parallel to $\overleftrightarrow{BC}$ and r units above it. The construction follows this analysis.

Construct right triangle BCA', with $BC = a$, right angle at vertex C, and $m\angle CBA' = $ complement of α; then circumscribe this triangle. This circle, (O, OB), is the circumcircle of the solution triangle because any angle inscribed in $\overset{\frown}{BA'C}$ will have angle measure α, as arranged for $\angle BA'C$. Draw radius $\overline{OP}$ perpendicular to $\overline{BC}$ and then draw the circle (P, PB), which is our first locus, L_1, for the incenter I. Now draw the second locus, L_2, which is a line parallel to $\overleftrightarrow{BC}$ and r units above it. These two loci meet at the incenter I, and then $\overleftrightarrow{PI}$ meets the circumcircle at A, the third vertex of solution triangle ABC.

Discussion   Of course, L_1 and L_2 must meet if we are to locate I, and if they meet once, they may meet again. We leave further discussion about the number and nature of the solutions to the reader. ●

CONSTRUCTION 122 $\{h_a,\ h_b,\ r\}$

We will not solve this problem completely here but instead will indicate how we can reduce it to a problem we have already solved. Recall Theorem 7.13, which stated that $\dfrac{1}{r} = \dfrac{1}{h_a} + \dfrac{1}{h_b} + \dfrac{1}{h_c}$. The algebraic consequence of this equation is that given any three of these four quantities we can find the fourth. Because this problem starts with h_a, h_b, and r as given, it has now been reduced, theoretically at least, to Construction 16, in which we were given the three altitudes.

There still remains the question of constructing the reciprocal of any given length x. We require a unit length, PT, and any circle tangent to $\overleftrightarrow{PT}$ at point T (see Figure 9-24). Let the circle (P, x) intersect this circle at point Q and consider R, the other intersection of $\overleftrightarrow{PQ}$ with the first circle. Because $PQ \cdot PR = PT^2 = 1$, it follows that PQ and PR are reciprocals. If (P, x) does not intersect the first circle, just take any larger circle tangent to $\overline{PT}$ at point T and proceed as before.

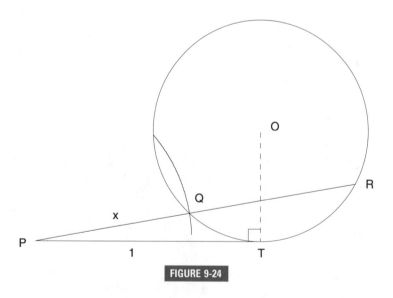

FIGURE 9-24

The actual solution construction will only be indicated here and not completely done. Use the reciprocal construction to find reciprocals of r, h_a and h_b. Subtract to find the reciprocal of h_c:

$$\frac{1}{h_c} = \frac{1}{r} - \frac{1}{h_a} - \frac{1}{h_b}$$

Then find the reciprocal of this reciprocal to obtain h_c itself. Now we are right back to Construction 16, which we did earlier (page 183).

CONSTRUCTION 150 $\{\alpha, h_a, s\}$

Suppose the solution triangle is available. On $\overleftrightarrow{BC}$, make $BP = BA = c$ and $CQ = CA = b$ (Figure 9-25). Thus the length of $\overline{PBCQ}$ is $a + b + c = 2s$, which is known. In isosceles triangle BPA, the measure of each of the congruent base angles is half the measure of the exterior angle at vertex B; that is, $m\angle PAB = \frac{1}{2}\beta$. Analogously, $m\angle QAC = \frac{1}{2}\gamma$. Thus, at vertex A, we have:

$$m\angle PAQ = \frac{1}{2}\beta + \alpha + \frac{1}{2}\gamma = \left(\frac{1}{2}\alpha + \frac{1}{2}\beta + \frac{1}{2}\gamma\right) + \frac{1}{2}\alpha = 90° + \frac{1}{2}\alpha$$

Thus $m\angle PAQ$ is also known in terms of the original given material.

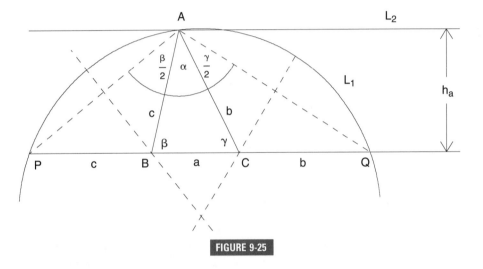

FIGURE 9-25

On the known segment $\overline{PQ}$, the point A subtends a known angle $90° + \frac{1}{2}\alpha$ and the locus of point A is thus known to be a circular arc, L_1. Another locus for point A is the line L_2, parallel to $\overleftrightarrow{BC}$ and at a distance h_a above it. We start the actual construction by finding L_1.

On any line, draw $PQ = 2s$, which is the perimeter of the solution triangle ABC. At point P, construct $\overrightarrow{PU}$ perpendicular to $\overleftrightarrow{PQ}$ and then make $\angle UPV$ congruent to the given angle of measure α (see Figure 9-26). Bisect $\angle UPV$ with $\overrightarrow{PW}$; thus $m\angle QPW = 90° + \frac{1}{2}\alpha$, as in the analysis. Then O, the center of arc L_1, will be found at the intersection of the perpendicular bisector of $\overline{PQ}$ and the perpendicular to $\overleftrightarrow{PW}$ at point P. Finally, (O, OP) is the actual locus L_1.

The other locus, L_2, is easily drawn, as indicated in Figure 9-26, and the two loci intersect at vertex A of the solution triangle. Then, at last, the perpendicular bisectors of $\overline{AP}$ and $\overline{AQ}$ meet the base line at points B and C, the other two vertices of the solution triangle ABC.

We have solved only a few of the 179 problems on our list, but we have selected those that lead to relevant and, we hope, interesting geometric material. We urge you to explore this territory in much more detail because you will be rewarded with knowledge, satisfaction, and pleasure.

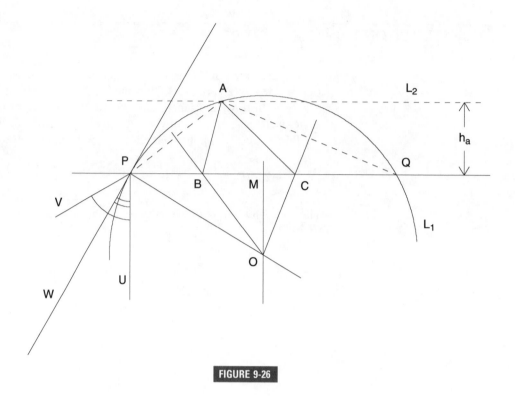

FIGURE 9-26

--------- EXERCISES ---------

1. Refer to Construction 20 (Figure 9-12). Could we take M_b as the other intersection of the two loci? Explain your answer.

2. Refer to Construction 29 (Figure 9-13). The arc $(C, 2m_c)$ may intersect $\overleftrightarrow{BD}$ at another point, D'. Discuss the construction and the solution you get from there on.

3. Refer to Construction 29 (Figure 9-13). If a has length 10 inches, what are possible lengths for h_b? For m_c? Discuss your response.

4. Refer to Construction 35 (Figure 9-14). Suppose we took K and H_a on the same side of $\overline{AM_a}$ rather than on opposite sides, as in our figure. Complete the new figure and discuss your findings.

5. Refer to Construction 35 (Figure 9-14). Under what conditions would $\overleftrightarrow{AK}$ and $\overrightarrow{H_aM_a}$ fail to meet? How would our "solution" be affected in that case?

6. Refer to Construction 35 (Figure 9-14). If $m_a = 10$ inches, discuss the possible lengths of h_a and h_b for any solution or for any number of solutions.

7. Refer to Construction 43 (Figure 9-15). What conditions must the lengths m_a, m_b, and m_c satisfy if we are to have any solution?

8. Refer to Construction 43 (Figure 9-15). What are the consequences if any two or all three of the given medians have equal length?

9. Refer to Construction 56 (Figure 9-16). What conditions on a and h_b would make it impossible to construct right triangle BCH_b?

10. Refer to Construction 56 (Figure 9-16). What length of the angle bisector $\overline{CT_c}$ would make it impossible for $\overleftrightarrow{BT_c}$ and $\overleftrightarrow{CH_b}$ to meet?

11. Refer to Construction 74 (Figure 9-18). Discuss the situations in which there is any equality among the three given lengths.

12. Refer to Construction 74 (Figure 9-18). We took M_a and T_a on the same side of H_a. Discuss the consequences of taking these points on opposite sides of H_a.

13. Refer to Construction 74 (Figure 9-18). Under what circumstances would the circle (O, OA) fail to meet $\overleftrightarrow{H_aM_a}$?

14. Refer to Construction 99 (Figure 9-19). Discuss the number and possibility of solutions with different selections of the given lengths $\{a, m_b, R\}$.

15. For Construction 99, see if you can arrive at the following necessary and sufficient conditions for any solution

$$\sqrt{R^2 + 2a^2} - R \leq 2m_b \leq \sqrt{R^2 + 2a^2} + R$$

16. Refer to Construction 102 (Figure 9-20). Under what circumstances would it be impossible to draw the first right triangle, $\triangle AH_aM_a$?

17. Refer to Construction 102 (Figure 9-20). Under what circumstances would the loci L_1 and L_2 fail to intersect?

18. Refer to Construction 102 (Figure 9-20). Loci L_1 and L_2 might intersect twice: at O, as shown, and at O', not shown. Follow through on this possibility.

19. Refer to Construction 102 (Figure 9-20). Once we have the intersection O, could we ever fail to get the last two vertices, B and C? Discuss your response.

Our last exercise in this chapter is a big one!

20. Complete the solutions of as many of the 179 problems on pages 179–180 as you can. Discuss for each the conditions for the possibility of any solution and the relations among conditions for the nature and number of any solutions.

CIRCLE
CONSTRUCTIONS

INTRODUCTION

In this chapter, we investigate in detail the solution of problems that involve constructing a circle to fit given conditions. One situation was the construction of a circle through the three vertices of a triangle, that is, the *circumscribed circle* of a given triangle ($\triangle ABC$). The solution to this problem is well known and unique: The perpendicular bisectors of the sides are concurrent at the circumcenter O, and the required circle has center O and radius $\overline{OA}$ (or $\overline{OB}$ or $\overline{OC}$).

Also previously discussed was the construction of the *inscribed circle* of a given triangle, that is, the circle tangent to the three sides of a triangle. This solution is also well known and also unique: The three angle bisectors are concurrent at the incenter I, which is the center of the desired circle. The radius is the distance from I to any of the three sides of the original triangle.

THE PROBLEM OF APOLLONIUS

Both of these problems involve the construction of a circle through given points or tangent to given lines. A natural generalization would be the problem of constructing a circle through one or more points (P) *and* tangent to one or more lines (L), *and* perhaps tangent to one or more circles (C). This larger, general problem, sometimes called the "problem of Apollonius,"* is analyzed for the ten situations listed here:

1. PPP	3. PLL	5. PPC	7. LLC	9. LCC
2. PPL	4. LLL	6. PLC	8. PCC	10. CCC

We will examine each of these cases in the constructions that follow.

---

* Apollonius (ca. 262 B.C.–ca. 190 B.C.) was born in Perga, a small Greek city in southern Asia Minor. The fame that Apollonius enjoys today results from his work on conic sections. In addition to his work *Conics,* Pappus mentions the contents of six other works of Apollonius, which form part of the *Treasury of Analysis.* The only work to have survived (originally written in Arabic and translated into Latin by Edmund Halley in 1706) is referred to as the "Two Books." It is his treatment of a tangencies problem, today referred to as the "problem of Apollonius," that we study in this chapter. Euclid, in *Elements,* Book IV, treated the first two cases. Apollonius's *Book I* treated cases 3, 4, 5, 6, 8, and 9, while the treatment of 7 and 10 took up all of *Book II.* Many famous mathematicians, notably Vieta and Newton, were fascinated by case 10. The reconstruction of the remaining four books of Apollonius consumed mathematicians in the seventeenth and eighteenth centuries.

▌CONSTRUCTION 1   PPP

We have discussed this case when the three points are the vertices of a triangle, but a special situation to be considered is the one in which the three points do not form a triangle, that is, they are collinear. In this event, the only "circle" to go through the three points is a "circle" of infinite radius, that is, a straight line.

▌CONSTRUCTION 2   PPL

If the solution were available, we would see that the line containing chord $\overline{P_1P_2}$ meets the given line at a point, A, that is an external point from which a tangent and secant are drawn to a circle (see Figure 10-1). But we know that in such a situation the tangent length is the mean proportional between the length of the whole secant and the length of its external segment.

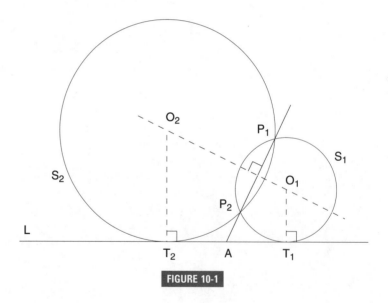

FIGURE 10-1

Let $\overleftrightarrow{P_1P_2}$ intersect line L at point A. Find, by construction, the mean proportion, t, between AP_1 and AP_2, and on line L, on either side of point A, draw points T_1 and T_2 so that $AT_1 = AT_2 = t$. Thus T_1 and T_2 are tangent points of the required circles on line L. The centers of these circles are on *both* the perpendicular bisector of $\overline{P_1P_2}$ *and* the perpendicular to L at points T_1 and T_2.

CONSTRUCTION 3   PLL

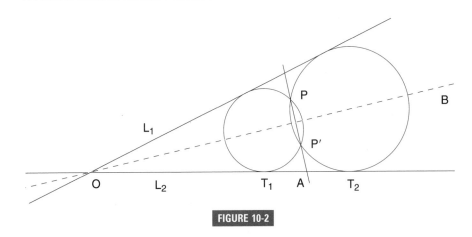

FIGURE 10-2

If the solution were available, we would see that there are, in general, two solutions with two common points P and P'. These points must be symmetrically placed with respect to the angle bisector $\overrightarrow{OB}$ (see Figure 10-2). Furthermore, the line containing common chord $\overline{PP'}$ must meet one of the sides, say L_2, at point A. We can find $\overrightarrow{OB}$, a bisector of one of the angles formed by the given lines L_1 and L_2, and we can thus also find P', the point symmetric to P with respect to $\overleftrightarrow{OB}$. Therefore we have reduced the case PLL to the case PPL discussed previously.

CONSTRUCTION 4   LLL

We have discussed the inscribed circle of the triangle formed by three lines that meet in three distinct points, A, B, and C. Figure 10-3 shows the situation that develops when we consider the three lines that form $\triangle ABC$. The constructions are simple enough: Because each circle is tangent to three lines, its center must lie on the bisectors of the angles formed by these lines. Once we have the centers, we can easily find the radii (how?) and then draw the required circles. We urge you to study this figure carefully; it shows some remarkable properties involving collinearity and perpendicularity.

CONSTRUCTION 5   PPC

Suppose the solution is available. Then the circles would be tangent at point T, at which point we could draw a common tangent line. If from any point A on this tangent line we draw secants to intersect the desired circle at points P_1 and P_2 and the given circle at points Q and R (see Figure 10-4), we would have

$$AP_1 \cdot AP_2 = AT^2 = AR \cdot AQ$$

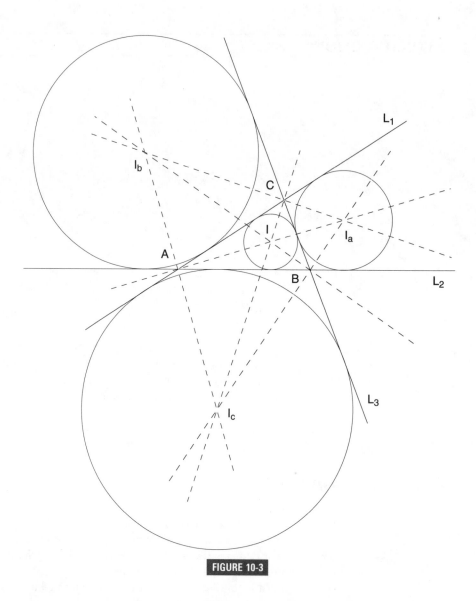

FIGURE 10-3

Thus, because $AP_1 \cdot AP_2 = AR \cdot AQ$, the four points P_1, P_2, Q, and R are cyclic. (See exercise 18 at the end of this chapter.) We can easily draw a circle through given points P_1 and P_2 that will intersect given circle C. The construction follows.

First draw the perpendicular bisector of $\overline{P_1P_2}$ (this is a locus for centers of all circles through P_1 and P_2). On this perpendicular, take any point E and draw the circle (E, EP_1) to intersect given circle C at points Q and R. Draw $\overleftrightarrow{QR}$ to intersect $\overleftrightarrow{P_1P_2}$ at point A. From point A, draw tangent $\overleftrightarrow{AT}$ to the given circle. Then draw $\overline{ST}$ through center S of the given circle and point T just found, to intersect the perpendicular bisector of $\overline{P_1P_2}$ at O, the center of the required circle. (The tangent $\overleftrightarrow{AT}$ to the given circle is only one of two possible tangent lines. Discuss the construction that uses the other tangent line, $\overline{AT'}$, not drawn in Figure 10-4.)

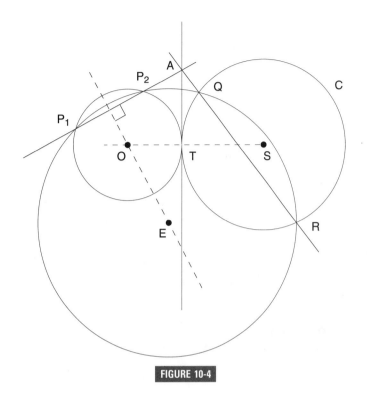

FIGURE 10-4

CONSTRUCTION 6   PLC

If the solution were available, it could appear in Figure 10-5 as circle
(E, EG), tangent to the given circle C at point T, tangent to the given line L at
point G, and passing through the given point P. The line of centers, $\overleftrightarrow{OE}$, must go
through point T (why?). Draw the perpendicular from center O to line L, intersecting the given circle at points A and B, the ends of a diameter; then draw the

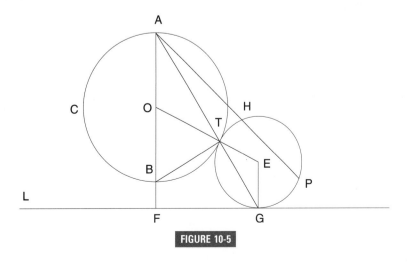

FIGURE 10-5

perpendicular from center E to line L, meeting it at point G, which is a point of tangency (why?). Then draw $\overline{BT}$, $\overline{AT}$, and $\overline{TG}$ and finally draw $\overline{AP}$ to intersect the desired circle at point H. $\overleftrightarrow{OF}$ and $\overleftrightarrow{EG}$ are parallel, and $\triangle OAT$ and $\triangle TEG$ are isosceles, with congruent vertex angles at centers O and E.

Thus the base angles of $\triangle OAT$ and $\triangle TEG$ are congruent, and in particular $\angle OAT \cong \angle ETG$. Hence $\overline{AT}$ and $\overline{TG}$ lie on the same straight line and $\angle BTG$ and $\angle BTA$ are right angles. Because $\angle BFG$ is also a right angle, we know that quadrilateral $BFGT$ is cyclic (on diameter $\overline{BG}$). Therefore we have secants from outside point A, and hence $AB \cdot AF = AT \cdot AG$. But points T, G, P, and H are also cyclic, and, as before, $AT \cdot AG = AH \cdot AP$; therefore $AB \cdot AF = AH \cdot AP$. Because points A, B, F, and P are all quickly available from the given material, we can find point H. The problem has thus been reduced to PPL, or Construction 2.

Briefly, to find point H we draw the perpendicular from O, the center of given circle C, to line L, intersecting circle C at points A and B and meeting line L at point F. Draw $\overleftrightarrow{AP}$, and on it construct $\overline{AH}$, found from $\dfrac{AP}{AF} = \dfrac{AB}{AH}$. Now proceed with P, H, and L as with $P_1 P_2 L$ in Construction 2.

CONSTRUCTION 7   LLC

Suppose, as usual, that a solution S is available, as shown in Figure 10-6. Because circle S is tangent to given circle C, the length of $\overline{OA}$, joining their centers, is equal to the sum of their radii, x unknown and r known and given. Thus another circle S', concentric with circle S and with radius $x + r$, will go through the center A of given circle C and be tangent to lines L_1' and L_2', parallel respectively to the given lines L_1 and L_2 and at distance r beyond them. But because these lines are easily constructed, the problem has been reduced to that of Construction 3 (PLL), with P the center A of given circle C and with the lines as L_1' and L_2' parallel to the given lines L_1 and L_2 and at distance r beyond them. The solution circle S to this problem will give us the desired center O. We know from

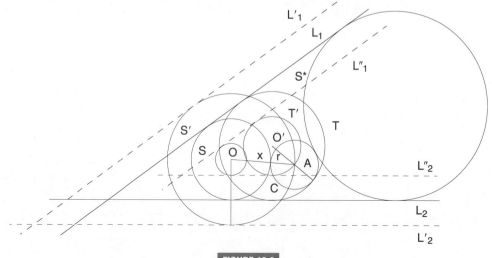

FIGURE 10-6

Construction 3 that there are, in general, two solutions, so we should have two solutions here also: S and $S^\star$, as sketched in Figure 10-6.

Another possibility must still be investigated: Suppose the solution circle T is *internally* tangent to the given circle. In that case, the length of $\overline{O'A}$, the segment joining their centers, will be equal to the *difference* of their radii rather than the sum, as before. Now a new auxiliary circle T', concentric with circle T but inside it and with radius $x' - r$, will go through the center A of the given circle and be tangent to two auxiliary lines L_1'' and L_2'', parallel respectively to given lines L_1 and L_2 but *inside* the angle formed by their intersection rather than beyond the angle as before. We have again reduced the problem to PLL, with point A and lines L_1'' and L_2''. There are two solutions in this case also, but only one of them, circle T, is shown.

CONSTRUCTION 8 PCC

Suppose that a solution, circle S, is available, tangent externally to both given circles, C_1 and C_2, at points T_1 and T_2, respectively, as in Figure 10-7. Let the common tangent $\overleftrightarrow{K_1K_2}$ to the given circles meet the line of centers, $\overleftrightarrow{O_1O_2}$, at point R. This is a center of similitude of these two circles. Assume that $\overleftrightarrow{T_1T_2}$ will go through point R (this is true; can you prove it?) and draw the other lines as indicated in Figure 10-7. Then $\triangle U_1O_1T_1$, $\triangle T_1OT_2$, and $\triangle T_2O_2U_2$ are all isosceles and *all* their base angles are congruent. Thus $\overleftrightarrow{O_1U_1} \parallel \overleftrightarrow{OO_2}$ and $\overleftrightarrow{OO_1} \parallel \overleftrightarrow{O_2U_2}$. Thus, from two pairs of similar triangles, we have:

$$\frac{RU_1}{RT_2} = \frac{RO_1}{RO_2} = \frac{RT_1}{RU_2}$$

Therefore $RU_1 \cdot RU_2 = RT_1 \cdot RT_2$. But for each circle separately, we have:

$$RU_1 \cdot RT_1 = RK_1^2 \quad \text{and} \quad RU_2 \cdot RT_2 = RK_2^2$$

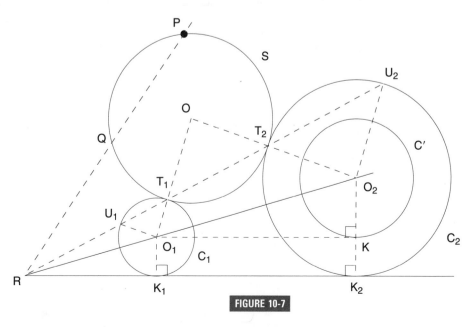

FIGURE 10-7

Thus:

$$(RU_1 \cdot RT_1) \cdot (RU_2 \cdot RT_2) = RK_1^2 \cdot RK_2^2 = (RU_1 \cdot RU_2) \cdot (RT_1 \cdot RT_2)$$
$$= (RT_1 \cdot RT_2)^2$$

Therefore $RT_1 \cdot RT_2 = RK_1 \cdot RK_2$. This means that the four points T_1, T_2, K_2, and K_1 are cyclic (as are the four points T_1, T_2, P, and Q), as indicated in exercise 18. Thus $RQ \cdot RP = RT_1 \cdot RT_2 = RK_1 \cdot RK_2$.

This last product is known because it is obtainable directly from the given circles as soon as we construct their common external tangent. (To construct this tangent, we start by drawing the circle $C' = (O_2, O_2K)$, where the length of $\overline{O_2K}$ is equal to the difference of the radii of the given circles; then we draw a tangent line from O_1 to this circle C'. The common tangent to the two circles will be parallel to this tangent $\overleftrightarrow{O_1K}$ and at distance O_1K_1 beyond it.)

From $RQ \cdot RP = RK_1 \cdot RK_2$, we locate point Q in $\overleftrightarrow{RP}$ by a fourth proportional construction and have thus reduced the problem to Construction 5 (PPC), with points P and Q and either of the given circles.

Because circle S could have been drawn internally tangent to either or both of the given circles and because P might be in a variety of locations relative to the given circles, which might in turn be in a variety of positions relative to each other, there are many specific situations to investigate. We leave the details to the reader.

CONSTRUCTION 9   LCC

We suppose, as usual, that a solution is available, as depicted in Figure 10-8. We are given circle C_1 and circle C_2 with radii r_1 and r_2, respectively, and line L. If we draw the circle S' concentric with the solution circle S but with radius $\overline{OO_1}$, then we have "expanded" the solution circle S to S'. S' will pass through O_1, be tangent to a new circle C_2' that is concentric with circle C_2 with radius equal to the difference of

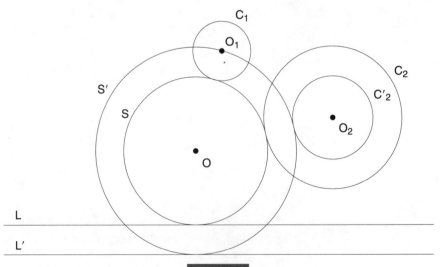

FIGURE 10-8

the given radii, and be tangent to a new line L' that is parallel to line L and beyond it by the distance r_1. Clearly we can construct circle C_2' and line L'; thus we can construct circle S' as a solution of Construction 6 (PLC) with given point O_1, given line L', and given circle C'. Once we have circle S', it is easy to get the solution circle S.

CONSTRUCTION 10   CCC

This last construction of the set is often called by itself the "problem of Apollonius" or the "circle of Apollonius." The given circles may lie in various relative positions, any one of which may lead to a number of solutions. (Can you draw the given circles so that there is *no* solution?)

We discuss here the most general case, that in which the circles are exterior to one another. This case leads, in general, to eight solutions. We draw only one solution—the circle that is externally tangent to all three of the given circles. Other solutions would be externally tangent to some of the given circles and internally tangent to the others. Suppose the solution is available, as usual. Circle S with radius r is to be drawn tangent to given circles C_1, C_2, and C_3, with centers O_1, O_2, and O_3 and radii r_1, r_2, and r_3 (see Figure 10-9).

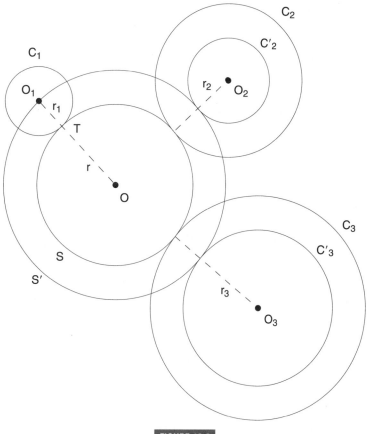

FIGURE 10-9

With the solution to Construction 9 in mind, we can "expand" the solution circle S concentrically to circle S' with radius $r + r_1$. Then we "shrink" circle C_1 to its center O_1, circle C_2 to circle C_2' with radius $r_2 - r_1$, and circle C_3 to circle C_3' with radius $r_3 - r_1$. Thus circle S' would pass through point O_1 and be tangent to circles C_2' and C_3', which brings us neatly back to Construction 8 (PCC) because we have point O_1 and can easily construct circles C_2' and C_3'. We easily get the solution circle S by "shrinking" circle S' concentrically, with center O and radius $OT = OO_1 - r_1$.

We've sketched another solution, in which circle S is internally tangent to circle C_1 but externally tangent to circles C_2 and C_3, as shown in Figure 10-10. The solution proceeds very much as before, with circle S' still found as a solution to the construction PCC, but in this case circle C_2' has radius $r_2 + r_1$ and circle C_3' has radius $r_3 + r_1$.

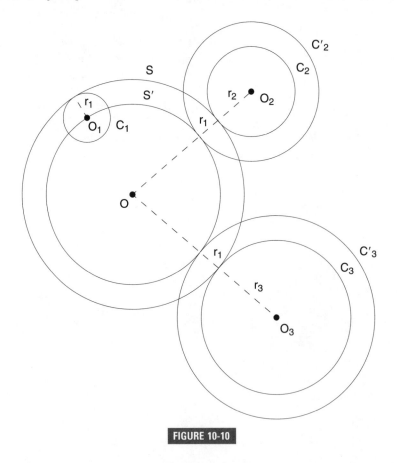

FIGURE 10-10

This completes our discussion of all ten construction problems, but you would be missing a lot of interesting geometry if you did not try to clear up all the delightful special cases that arise when we move the given parts around a bit, as in the exercises that follow.

———————————— E XERCISE S ————————————

We indicate special situations that may exist among the given parameters in each construction. These special situations lead to different solutions and different numbers of solutions.

For exercises 1–4 (PPL), discuss the situation for the special cases indicated.

1. $\overleftrightarrow{P_1 P_2}$ is parallel to L (one solution).

2. P_1 is on L (one solution).

3. P_1 and P_2 are both on L.

4. L lies between P_1 and P_2.

For exercises 5–9 (PLL), discuss in detail the special cases indicated.

5. L_1 and L_2 are parallel, and P lies on L_1.

6. L_1 and L_2 are parallel, and P lies between them.

7. L_1 and L_2 are parallel, and P lies outside of them.

8. L_1 and L_2 intersect, and P lies on L_1.

9. L_1 and L_2 intersect at P.

Exercises 10–17 (LLL, Figure 10-3) indicate some details that you are asked to work out by yourself. Try to find (and prove) other relations.

10. The circle with center I is called the *inscribed* circle of $\triangle ABC$ and has radius r. The circles with centers I_a, I_b, and I_c are called the *escribed* circles of $\triangle ABC$, with radii r_a, r_b, and r_c, respectively. They are all related by a remarkable formula:

$$\frac{1}{r} = \frac{1}{r_a} + \frac{1}{r_b} + \frac{1}{r_c}$$

Prove this formula. (Hint: Use areas.)

11. The four radii of the inscribed and escribed circles are also related to $\triangle ABC$ in other ways: Each can be found directly from the lengths of the sides. Thus, with semiperimeter $s = \frac{1}{2}(a + b + c)$, we have:

$$r = \sqrt{\frac{(s - a)(s - b)(s - c)}{s}} \qquad r_a = \sqrt{\frac{s(s - b)(s - c)}{(s - a)}}$$

$$r_b = \sqrt{\frac{s(s - a)(s - c)}{(s - b)}} \qquad r_c = \sqrt{\frac{s(s - a)(s - b)}{(s - c)}}$$

Prove these formulas. (Hint: Use areas, particularly Heron's formula:
area $\triangle ABC = \sqrt{s(s-a)(s-b)(s-c)}$.)*

12. Prove that the reciprocal of the inradius is equal to the sum of the reciprocals of the lengths of the altitudes, that is, $\dfrac{1}{r} = \dfrac{1}{h_a} + \dfrac{1}{h_b} + \dfrac{1}{h_c}$.

13. a. Prove that the product of all four of the radii is equal to the square of the area of the triangle.
    b. Prove that the sum of the three exradii is equal to the sum of the inradius and four times the circumradius.

14. Prove that each vertex of $\triangle ABC$ is collinear with the incenter, I, and its opposite excenter (e.g., A, I, and I_a are collinear).

15. Prove that each vertex of $\triangle ABC$ is also collinear with its two adjacent excenters (e.g., A, I_b, and I_c are collinear).

16. Prove that in $\triangle I_a I_b I_c$ the altitudes are exactly the bisectors of the angles of $\triangle ABC$, which meet at the incenter, I.

17. Using the result from exercise 16, prove that the four points I, I_a, I_b, and I_c form an orthic quadrilateral, which means that if we select any three of them and draw the altitudes of that triangle, then those altitudes will be concurrent at the fourth point.

For exercises 18 and 19 (PPC), prove the relationships indicated.

18. If, as indicated, $\overleftrightarrow{P_1 P_2}$ and $\overleftrightarrow{QR}$ intersect at point A and $AP_1 \cdot AP_2 = AR \cdot AQ$, prove that P_1, P_2, Q, and R are cyclic. (Hint: Obtain a proportion from the given equation, then prove a pair of triangles similar, then get a pair of angles supplementary, and then use the fact that a quadrilateral can be inscribed in a circle if and only if both pairs of opposite angles are supplementary.)

19. Prove that if two circles are externally tangent, their line of centers contains their common point of tangency.

Discuss each of the special situations in exercises 20–31 (LLC), with figures and constructions.

20. L_1 and L_2 are parallel, and C is tangent to both.

21. L_1 and L_2 are parallel, and C intersects both.

22. L_1 and L_2 are parallel, and C intersects L_1 and is tangent to L_2.

23. L_1 and L_2 are parallel, and C intersects L_1 but not L_2.

24. L_1 and L_2 are parallel, and C lies between them and is tangent to L_1 but not to L_2.

---

* For a proof of Heron's formula, see Alfred S. Posamentier and Charles T. Salkind, *Challenging Problems in Geometry* (New York: Dover, 1988), pp. 135–137.

25. L_1 and L_2 are parallel, and C lies between them but is not tangent to either.

26. L_1 and L_2 intersect at point K, which is interior to C.

27. L_1 and L_2 intersect at point K, which is on C.

28. L_1 and L_2 intersect at point K, and C is tangent to L_1 at point K.

29. L_1 and L_2 intersect at point K outside C, but C is tangent to L_1 and L_2.

30. L_1 and L_2 intersect at point K outside C, but C is tangent to L_1 and intersects L_2.

31. L_1 and L_2 intersect at point K outside C, but C intersects both L_1 and L_2.

Discuss each of the special situations in exercises 32–39 (LCC), with figures and constructions.

32. C_1 is inside C_2 (not tangent), and L intersects both C_1 and C_2.

33. C_1 is inside C_2 (not tangent), and L is tangent to C_1.

34. C_1 is inside C_2 (not tangent), and L intersects C_2 but not C_1.

35. C_1 is inside C_2 (not tangent), and L is tangent to C_2.

36. C_1 is inside C_2 (not tangent), and L intersects neither circle.

37. C_1 is internally tangent to C_2 at point T. Discuss the solutions for each possible position of L relative to these circles as described in exercises 32–36.

38. C_1 and C_2 intersect at points P and Q. Discuss the nature and number of solutions for each position of L relative to these two circles as described in exercises 32–36.

39. C_1 and C_2 are externally tangent at point T. Discuss the nature and number of solutions for each position of L relative to these two circles as described in exercises 32–36.

For each situation in exercises 40–59 (CCC), draw the figure and discuss and carry out the constructive solution.

40. C_1 is inside C_2 (not tangent), which is inside C_3 (not tangent).

41. C_1 is inside C_2 (not tangent), which is internally tangent to C_3.

42. C_1 is inside C_2 (not tangent), and C_3 intersects both C_1 and C_2.

43. C_1 is inside C_2 (not tangent), and C_3 intersects C_2 but not C_1.

44. C_1 is inside C_2 (not tangent), and C_3 intersects C_2 and is tangent to C_1.

45. C_1 is inside C_2 (not tangent), and C_3 is tangent to both C_1 and C_2.

46. C_1 is inside C_2 (not tangent), and C_3 intersects C_1 but not C_2.

47. C_1 is inside C_2 (not tangent), and C_3 is externally tangent to C_2.

48. C_1 is inside C_2 (not tangent), and C_3 is also inside C_2 (not tangent) but outside C_1.

49. C_1 is inside C_2 (not tangent), C_3 is also inside C_2 (not tangent), and C_1 and C_3 are externally tangent.

50. C_1 is internally tangent to C_2 at point T_1, and C_3 is tangent to both C_1 and C_2 at point T_1 (four cases).

51. C_1 is internally tangent to C_2 at point T_1, and C_3 is tangent to both C_1 and C_2 but not at point T_1.

52. C_1 is internally tangent to C_2 at point T_1, and C_3 intersects both C_1 and C_2 at point T_1 as well as at other points.

53. C_1 is internally tangent to C_2 at point T_1, and C_3 intersects both C_1 and C_2 but not at point T_1.

54. C_1 is internally tangent to C_2 at point T_1, and C_2 is internally tangent to C_3 at point T_2.

55. C_1 is internally tangent to C_2 at point T_1, and C_3 is internally tangent to C_2 but not at point T_1 and intersects C_1.

56. C_1 is internally tangent to C_2 at point T_1, and C_3 is tangent to C_1 and intersects C_2.

57. C_1 is internally tangent to C_2 at point T_1, and C_3 is externally tangent to C_2 at point T_2.

58. C_1 is internally tangent to C_2 at point T_1, and C_3 is exterior to both C_1 and C_2.

59. C_1 intersects C_2 at points P and Q. Discuss the possible positions of C_3 relative to the given circles and the consequent solutions.

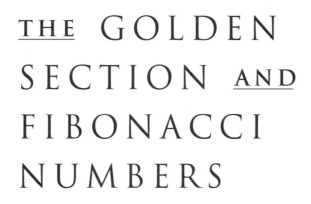

THE GOLDEN SECTION AND FIBONACCI NUMBERS

THE GOLDEN RATIO

To this point our excursion into advanced Euclidean geometry has been largely geometrical. Yet geometry interrelates with many other branches of mathematics. It is therefore appropriate for our last chapter to exhibit an example of this inter-relationship. In this chapter, we show some astonishing relationships between the golden section (or golden ratio) and the Fibonacci numbers. Many interesting relationships will be unveiled as we study these topics.

We begin by introducing and defining the golden section, or, as it is often called, the *golden ratio*. This ratio, $\frac{a}{b} \approx 1.61803398874987\ldots$, at first sight seems like nothing special. You will soon see that it turns into quite an interesting ratio. You will find that $\frac{b}{a} \approx 0.61803398874987\ldots$. When have we previously seen that $\frac{a}{b} = \frac{b}{a} + 1$? This is quite unusual! So read on and let yourself be entertained by this most unusual series of relationships.

FIGURE 11-1

Consider a point P located on $\overline{AB}$ so that $\frac{AB}{AP} = \frac{AP}{PB}$ (see Figure 11-1). We say that point P has partitioned $\overline{AB}$ in the *golden ratio*. Why is this ratio referred to as the *golden* ratio? Let us construct a rectangle whose length and width are the two segments $\overline{AP} = \ell$ and $\overline{BP} = w$. It is said that the shape of this rectangle is the most pleasing to look at. Through the ages, this rectangle has been associated with beauty. Which of the two rectangles in Figures 11-2 and 11-3 appears more pleasing to look at?

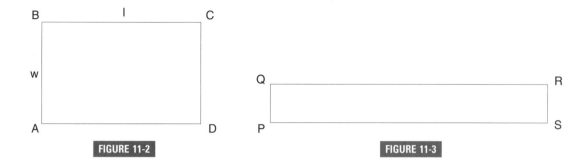

FIGURE 11-2   FIGURE 11-3

According to psychologists who have experimented with this question (Gustav Fechner, 1876 and Edward Lee Thorndike, 1917), the rectangle in Figure 11-2 is clearly more pleasing to look at. They indicate that a glance at rectangle *ABCD* catches the entire rectangle at once, while looking at rectangle *PQRS* (Figure 11-3) requires a horizontal scanning motion of the eyes by most people.

The beauty of this rectangle, called the *golden rectangle,* is not a new discovery. Ancient civilizations were quite familiar with it. For example, in architecture we find this rectangle in famous structures such as the Parthenon in Athens, Greece (Figure 11-4), and the doors of the Cathedral of Chartres, France (Figure 11-5). There are many examples of this type of rectangle around us. Try to find some. The golden rectangle has its length and width in a golden ratio; that is, $\dfrac{w}{\ell} = \dfrac{\ell}{w + \ell}$.

FIGURE 11-4

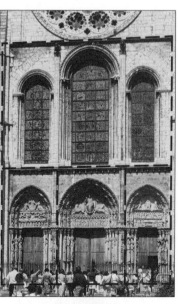

FIGURE 11-5

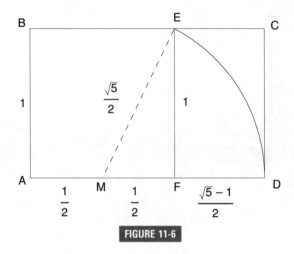

FIGURE 11-6

How do we construct a golden rectangle? Perhaps one of the simpler ways is to begin with a square, *ABEF* (see Figure 11-6), with *M* the midpoint of $\overline{AF}$. Then, with radius $\overline{ME}$ and center *M*, draw a circle to intersect $\overrightarrow{AF}$ at point *D*. The perpendicular to $\overrightarrow{AF}$ at point *D* intersects $\overrightarrow{BE}$ at point *C*. We now have rectangle *ABCD*, a golden rectangle.

Let us verify that rectangle *ABCD* is, in fact, a golden rectangle. Without loss of generality, we let quadrilateral *ABEF* be a unit square. Therefore $EF = AF = 1$ and $MF = \dfrac{1}{2}$. By the Pythagorean theorem, $ME = \dfrac{\sqrt{5}}{2}$.

Therefore $AD = \dfrac{\sqrt{5} + 1}{2}$. To verify that rectangle *ABCD* is a golden rectangle, we must show that:

$$\frac{w}{\ell} = \frac{\ell}{w + \ell} \quad \text{or} \quad \frac{CD}{AD} = \frac{AD}{CD + AD}$$

Substituting the above lengths, we get:

$$\frac{1}{\dfrac{\sqrt{5} + 1}{2}} = \frac{\dfrac{\sqrt{5} + 1}{2}}{1 + \dfrac{\sqrt{5} + 1}{2}}$$

This is a true equality.

The Greek letter ϕ is often used to represent the golden ratio. We can find an approximate value for ϕ (you might want to use a calculator!):

$$\phi = \frac{AD}{CD} = \frac{\sqrt{5} + 1}{2} \approx 1.61803398874987483\ldots$$

This number ϕ is most peculiar. Consider its reciprocal, $\dfrac{1}{\phi}$:

$$\frac{1}{\phi} = \frac{2}{\sqrt{5}+1} = \frac{\sqrt{5}-1}{2} = FD \approx 0.61803398874987483\ldots$$

Therefore not only is it true that $\phi \cdot \dfrac{1}{\phi} = 1$ (obviously!) but also $\phi - \dfrac{1}{\phi} = 1$ (try to verify this). Furthermore, ϕ and $\dfrac{1}{\phi}$ are the roots of the equation $x^2 - x - 1 = 0$, a property we will discuss later.

It is interesting to inspect powers of ϕ. We first must find the value of ϕ^2 in terms of ϕ:

$$\phi^2 = \left(\frac{\sqrt{5}+1}{2}\right)^2 = \frac{\sqrt{5}+3}{2} = \frac{\sqrt{5}+1}{2} + 1 = \phi + 1$$

The rest of the powers of ϕ can be obtained as follows:

$$\phi^3 = \phi \cdot \phi^2 = \phi(\phi + 1) = \phi^2 + \phi = (\phi + 1) + \phi = 2\phi + 1$$
$$\phi^4 = \phi^2 \cdot \phi^2 = (\phi + 1)(\phi + 1) = \phi^2 + 2\phi + 1 = (\phi + 1) + 2\phi + 1$$
$$= 3\phi + 2$$
$$\phi^5 = \phi^3 \cdot \phi^2 = (2\phi + 1)(\phi + 1) = 2\phi^2 + 3\phi + 1 = 2(\phi + 1) + 3\phi + 1$$
$$= 5\phi + 3$$
$$\phi^6 = \phi^3 \cdot \phi^3 = (2\phi + 1)(2\phi + 1) = 4\phi^2 + 4\phi + 1 = 4(\phi + 1) + 4\phi + 1$$
$$= 8\phi + 5$$
$$\phi^7 = \phi^4 \cdot \phi^3 = (3\phi + 2)(2\phi + 1) = 6\phi^2 + 7\phi + 2 = 6(\phi + 1) + 7\phi + 2$$
$$= 13\phi + 8$$
$$\vdots$$

$\phi = 1\phi + 0$
$\phi^2 = 1\phi + 1$
$\phi^3 = 2\phi + 1$
$\phi^4 = 3\phi + 2$
$\phi^5 = 5\phi + 3$
$\phi^6 = 8\phi + 5$
$\phi^7 = 13\phi + 8$
$\phi^8 = 21\phi + 13$
$\phi^9 = 34\phi + 21$
$\phi^{10} = 55\phi + 34$
$\vdots$

FIGURE 11-7

We summarize the powers of ϕ in Figure 11-7. Notice that the coefficients and constants seem somewhat related. As a matter of fact, they form a pattern that will become more familiar to you shortly. These numbers may be considered the basis for the connecting link between many branches of mathematics (including geometry). Before continuing with our study of the golden ratio in geometry, we will digress a bit to examine these numbers more carefully.

FIBONACCI NUMBERS

The origin of this sequence of numbers is rather interesting. The numbers first appeared as the solution to a problem posed in a book on algebraic methods and problems. This book, *Liber Abaci,* was written in 1202 by Leonardo of Pisa, better known as Fibonacci (1180–1250).* The problem we will examine involves the regeneration of rabbits. It may be stated as follows:

> How many pairs of rabbits will be produced in a year, beginning with a single pair, if in every month each pair bears a new pair, which becomes productive from the second month on?

It is from this problem that the famous *Fibonacci sequence* emerged. If we assume that a pair of baby rabbits (*B*) matures in one month to become a pair of offspring-producing adults (*A*), then we can set up the chart in Figure 11-8. The number of pairs of mature rabbits living each month determines the Fibonacci sequence (column 3):

$$1, 1, 2, 3, 5, 8, 13, 21, 34, 55, 89, 144, 233, 377, \ldots$$

If we let f_n be the nth term of the Fibonacci sequence, then:

$$f_1 = 1$$
$$f_2 = 1$$
$$f_3 = f_2 + f_1 = 1 + 1 = 2$$
$$f_4 = f_3 + f_2 = 2 + 1 = 3$$
$$f_5 = f_4 + f_3 = 3 + 2 = 5$$
$$\vdots \qquad \vdots$$
$$f_n = f_{n-1} + f_{n-2} \quad \text{for } n \text{ an integer} \geq 3$$

That is, each term after the first two terms is the sum of the two preceding terms.

---

* Fibonacci was not a clergyman, as might be expected of early scientists; rather, he was a merchant who traveled extensively throughout the Islamic world and took advantage of reading all he could of the Arabic mathematical writings. He was the first to introduce the Hindu-Arabic numerals to the Christian world in his *Liber Abaci* (1202; revised in 1228), which first circulated widely in manuscript form and was first published in 1857 as *Scritti di Leonardo Pisano* (Rome: B. Buoncompagni). The book is a collection of business mathematics, including linear and quadratic equations, square roots and cube roots, and other new topics, seen from the European viewpoint. He begins the book with: "These are the nine figures of the Indians 9 8 7 6 5 4 3 2 1. With these nine figures, and with the sign 0, which in Arabic is called *zephirum,* any number can be written, as will be demonstrated below." From here on he introduces the decimal position system for the first time in Europe. (Note: The word *zephirum* evolves from the Arabic word *as-sifr,* which comes from the Sanskrit word, used in India as early as the fifth century, *sunya,* referring to "empty.")

Month	Pairs	Number of Pairs of Adults (A)	Number of Pairs of Babies (B)	Total Pairs
Jan. 1	A	1	0	1
Feb. 1	A B	1	1	2
Mar. 1	A B A	2	1	3
Apr. 1	A B A A B	3	2	5
May 1	A B A A B A B A	5	3	8
June 1	A B A A B A B A A B A A B	8	5	13
July 1		13	8	21
Aug. 1		21	13	34
Sept. 1		34	21	55
Oct. 1		55	34	89
Nov. 1		89	55	144
Dec. 1		144	89	233
Jan. 1		233	144	377

FIGURE 11-8

Example   Using the "rule" of the Fibonacci sequence, find the ten numbers that would *precede* the first 1 if we allow nonpositive n.

Solution   Because $f_{n+1} = f_n + f_{n-1}$, we have $f_{n-1} = f_{n+1} - f_n$. Therefore:

$$f_0 = f_2 - f_1 = 1 - 1 = 0$$
$$f_{-1} = f_1 - f_0 = 1 - 0 = 1$$
$$f_{-2} = f_0 - f_{-1} = 0 - 1 = -1$$
$$f_{-3} = f_{-1} - f_{-2} = 1 - (-1) = 2$$
$$f_{-4} = f_{-2} - f_{-3} = -1 - 2 = -3$$
$$f_{-5} = f_{-3} - f_{-4} = 2 - (-3) = 5$$
$$f_{-6} = f_{-4} - f_{-5} = -3 - 5 = -8$$
$$f_{-7} = f_{-5} - f_{-6} = 5 - (-8) = 13$$
$$f_{-8} = f_{-6} - f_{-7} = -8 - 13 = -21$$
$$f_{-9} = f_{-7} - f_{-8} = 13 - (-21) = 34 \; \bullet$$

The preceding example was done deductively. Often, more pleasure is derived from an inductive discovery of a relationship, as in the next example.

Example   Find the sum of the first two terms, three terms, four terms, five terms, . . . , nine terms, and ten terms of the Fibonacci sequence. Then generalize this pattern to find the sum of the first n terms.

Solution   By taking the successive sums required $\left(\text{i.e., } \sum_{m=1}^{n} f_m \right)$, we may easily form the first three columns of the chart in Figure 11-9. Thus we have the sum of the first two Fibonacci numbers, $\sum_{m=1}^{2} f_m = f_1 + f_2 = 1 + 1 = 2$, the sum of the first three, $\sum_{m=1}^{3} f_m = f_1 + f_2 + f_3 = 1 + 1 + 2 = 4$, and so on. ●

m	f_m	$\sum_{m=1}^{n} f_m$	$\sum_{m=1}^{n} f_{2m-1}$	$\sum_{m=1}^{n} f_{2m}$	f_m^2	$\sum_{m=1}^{n} f_m^2$
1	1	1	1		1	1
2	1	2		1	1	2
3	2	4	3		4	6
4	3	7		4	9	15
5	5	12	8		25	40
6	8	20		12	64	104
7	13	33	21		169	273
8	21	54		33	441	714
9	34	88	55		1,156	1,870
10	55	143		88	3,025	4,895
11	89	232	144		7,921	12,816
12	144	376		132	20,736	33,552
13	233	609	377		54,289	87,841
14	377	986		609	142,129	229,970
15	610	1,596	987		372,100	602,070
16	987	2,583		1,596	974,169	1,576,239
17	1,597	4,180	2,584		2,550,409	4,126,648
18	2,584	6,764		4,180	6,677,056	10,803,704

FIGURE 11-9

Clearly, each term in the column of sums $\left(\sum_{m=1}^{n} f_m \right)$ is one less than a Fibonacci number. More specifically, $\sum_{m=1}^{n} f_m = f_{n+2} - 1$. That is, the sum of the first n Fibonacci numbers is one less than the $(n + 2)$nd Fibonacci number. We state this observation as Theorem 11.1.

THEOREM 11.1 $\sum_{m=1}^{n} f_m = f_{n+2} - 1$

Proof By the definition of Fibonacci numbers, $f_1 + f_2 = f_3$. Therefore:

$$\begin{aligned} f_1 &= \cancel{f_3} - f_2 \\ f_2 &= \cancel{f_4} - \cancel{f_3} \\ f_3 &= \cancel{f_5} - \cancel{f_4} \\ f_4 &= \cancel{f_6} - \cancel{f_5} \\ &\vdots \\ f_{n-1} &= \cancel{f_{n+1}} - \cancel{f_n} \\ f_n &= f_{n+2} - \cancel{f_{n+1}} \end{aligned}$$

By addition, $\sum_{m=1}^{n} f_m = f_{n+2} - f_2 = f_{n+2} - 1$. ●

Consider the series of odd-numbered terms of the Fibonacci sequence. Figure 11-9 can be helpful in finding the pattern of sums of these terms:

$$f_1 + f_3 + f_5 + f_7 + \cdots + f_{2m-1} = 1 + 2 + 5 + 13 + \cdots + f_{2m-1}$$

You may already have anticipated our next theorem.

THEOREM 11.2 $\sum_{m=1}^{n} f_{2m-1} = f_{2n}$

Proof

$$\begin{aligned} f_1 &= \cancel{f_2} \\ f_3 &= \cancel{f_4} - \cancel{f_2} \\ f_5 &= \cancel{f_6} - \cancel{f_4} \\ f_7 &= \cancel{f_8} - \cancel{f_6} \\ &\vdots \\ f_{2n-3} &= \cancel{f_{2n-2}} - \cancel{f_{2n-4}} \\ f_{2n-1} &= f_{2n} - \cancel{f_{2n-2}} \end{aligned}$$

By addition, $\sum_{m=1}^{n} f_{2m-1} = f_{2n}$. ●

We already know from Theorem 11.1 that:

$$\sum_{m=1}^{n} f_m = f_{n+2} - 1 \qquad \text{(I)}$$

If we double the number of terms in (I), we get:

$$\sum_{m=1}^{2n} f_m = f_{2n+2} - 1 \qquad \text{(II)}$$

Suppose we now subtract all the odd-numbered terms from (II); we would be left with the even-numbered terms. That is:

$$\begin{aligned}
\sum_{m=1}^{n} f_{2m} &= \sum_{m=1}^{2n} f_m - \sum_{m=1}^{n} f_{2m-1} \\
&= f_{2n+2} - 1 - f_{2n} \\
&= f_{2n+1} - 1 \quad (\text{because } f_{2n} = f_{2n+2} - f_{2n-1})
\end{aligned}$$

This result proves our next theorem.

THEOREM 11.3 $\displaystyle\sum_{m=1}^{n} f_{2m} = f_{2n+1} - 1$

An inspection of the f_m^2 column of the table of Fibonacci numbers in Figure 11-9 reveals a relationship between Fibonacci numbers and their squares. This relationship is stated as our next theorem.

THEOREM 11.4 $f_n^2 - f_{n-1} \cdot f_{n+1} = (-1)^{n-1}$

Before proving this theorem, we will discuss an alternate type of proof, *mathematical induction.* Consider a set of dominoes with an endless number of tiles set up as illustrated in Figure 11-10. If we were asked to knock down all the tiles, we could consider two methods: (1) We could knock down each tile separately, or (2) we could knock down just the first tile if we were sure that any tile that is knocked down automatically knocks down the tile after it.

The first method would not only be inefficient but would also never assure us of knocking down all the tiles (because the end might never be reached). The

FIGURE 11-10

second method guarantees that all the tiles are knocked down. After we have knocked down the first tile, we are then assured that any knocked down tile also knocks down its successor tile. That is, the first tile knocks down the second, which knocks down the third, which knocks down the fourth, and so on. All the tiles will be knocked down. This second method is directly analogous to the *axiom of mathematical induction:*

> A proposition involving the natural number n is true for all natural numbers:
>
> a. if the proposition is correct when $n = 1$;
> b. if given that the proposition is correct for $n = k$, the proposition is also correct for $n = k + 1$.

We now use mathematical induction to prove Theorem 11.4.

P roof   When $n = 1$, we have $f_n^2 = f_{n-1} \cdot f_{n+1} = f_1^2 - f_0 \cdot f_2 = 1 - (0)(1) = 1 = (-1)^{1-1}$ ($f_0 = 0$; see the first example on page 220).
We assume the proposition is true for $n = k$; that is:

$$f_k^2 - f_{k-1} \cdot f_{k+1} = (-1)^{k-1}$$

Now we show that the proposition is true for $n = k + 1$; that is:

$$f_{k+1}^2 - f_k \cdot f_{k+2} = (-1)^k$$

Because $f_{k+2} = f_k + f_{k+1}$:

$$\begin{aligned}
f_{k+1}^2 - f_k f_{k+2} &= f_{k+1}^2 - f_k(f_k + f_{k+1}) \\
&= f_{k+1}^2 - f_k^2 - f_k \cdot f_{k+1} \\
&= f_{k+1}(f_{k+1} - f_k) - f_k^2 \\
&= f_{k+1}f_{k-1} - f_k^2 \\
&= -(-1)^{k-1} \\
&= (-1)^k
\end{aligned}$$

This proves our theorem. ●

From Figure 11-9, you can see that the relationship that we state as our next theorem is justified.

THEOREM 11.5 $f_n \cdot f_{n+1} - f_{n-1} \cdot f_n = f_n^2$

P roof   (by mathematical induction) When $n = 1$, we have $f_1 \cdot f_2 - f_0 \cdot f_1 = 1 \cdot 1 - 0 \cdot 1 = 1 = f_1^2$.
We assume the proposition is true for $n = k$; that is:

$$f_k \cdot f_{k+1} - f_{k-1} \cdot f_k = f_k^2$$

We now show that the proposition is true for $n = k + 1$; that is:

$$f_{k+1} \cdot f_{k+2} - f_k \cdot f_{k+1} = f_{k+1}(f_{k+2} - f_k)$$

We know that $f_{k+2} - f_k = f_{k+1}$. Therefore:

$$f_{k+1} \cdot f_{k+2} - f_k \cdot f_{k+1} = f_{k+1}(f_{k+2} - f_k) = f_{k+1} \cdot f_{k+1} = f_{k+1}^2$$

This proves our theorem. ●

A careful inspection of $\displaystyle\sum_{m=1}^{n} f_m^2$ in Figure 11-9 reveals an interesting relationship: Each term equals the product of the corresponding Fibonacci number and the succeeding Fibonacci number. We state this relationship as our next theorem.

▌THEOREM 11.6 $\displaystyle\sum_{m=1}^{n} f_m^2 = f_n \cdot f_{n+1}$

℗roof From Theorem 11.5, we have $f_m^2 = f_m \cdot f_{m+1} - f_{m-1} \cdot f_m$. Therefore:

$$f_1^2 = f_1 \cdot f_2 - f_0 \cdot f_1 = \cancel{f_1 \cdot f_2} \quad \text{(because } f_0 = 0)$$
$$f_2^2 = \cancel{f_2 \cdot f_3} - \cancel{f_1 \cdot f_2}$$
$$f_3^2 = \cancel{f_3 \cdot f_4} - \cancel{f_2 \cdot f_3}$$
$$\vdots$$
$$f_{n-1}^2 = \cancel{f_{n-1} \cdot f_n} - \cancel{f_{n-2} \cdot f_{n-1}}$$
$$f_n^2 = f_n \cdot f_{n+1} - \cancel{f_{n-1} \cdot f_n}$$

By addition, $\displaystyle\sum_{m=1}^{n} f_m^2 = f_n f_{n+1}$. ●

There are many other fascinating relationships involving Fibonacci numbers. The proofs of some of them are offered as exercises.

LUCAS NUMBERS

Fibonacci numbers were so named in 1877 by the French mathematician François-Édouard-Anatole Lucas (1842–1891). At that time he also established another sequence of numbers, based on the same recursive definition but beginning with 1, 3 instead of 1, 1: the sequence 1, 3, 4, 7, 11, ... This sequence, which now bears the name Lucas numbers, is shown in Figure 11-11.

m	ℓ_m	$\sum_{m=1}^{n} \ell_m$	$\sum_{m=1}^{n} \ell_{2m-1}$	$\sum_{m=1}^{n} \ell_{2m}$	ℓ_m^2	$\sum_{m=1}^{n} \ell_m^2$
1	1	1	1		1	1
2	3	4		3	9	10
3	4	8	5		16	26
4	7	15		10	49	75
5	11	26	16		121	196
6	18	44		28	324	520
7	29	73	45		841	1,361
8	47	120		75	2,209	3,570
9	76	196	121		5,776	9,346
10	123	319		198	15,129	24,475
11	199	518	320		39,601	64,076
12	322	840		520	103,684	167,760
13	521	1,361	841		271,441	439,201
14	843	2,204		1,363	710,649	1,149,850
15	1,364	3,568	2,205		1,860,496	3,010,346
16	2,207	5,775		3,570	4,870,849	7,881,195
17	3,571	9,346	5,776		12,752,041	20,633,236
18	5,778	15,124		9,348	33,385,284	54,018,520

FIGURE 11-11

As was the case with the Fibonacci numbers, many interesting relationships can be discovered from the table in Figure 11-11. We will state some of these relationships as theorems.

THEOREM 11.7
$$\sum_{m=1}^{n} \ell_m = \ell_{n+2} - 3$$

P roof   We will use mathematical induction as an alternative to the method of proof used for Theorem 11.1.

Let $n = 1$; then $\sum_{m=1}^{1} \ell_m = \ell_{1+2} - 3 = \ell_3 - 3 = 4 - 3 = 1$.

We assume the proposition is true for $n = k$; that is:

$$\sum_{m=1}^{k} \ell_m = \ell_{k+2} - 3$$

We must now show that the proposition is true for $n = k + 1$; that is:

$$\sum_{m=1}^{k+1} \ell_m = \ell_{(k+1)+2} - 3 = \ell_{k+3} - 3$$

From our assumption, we have:

$$\sum_{m=1}^{k} \ell_m + \ell_{k+1} = \ell_{k+2} - 3 + \ell_{k+1}$$

$$\sum_{m=1}^{k+1} \ell_m = \ell_{k+3} - 3 \quad \text{(because } \ell_{k+2} + \ell_{k+1} = \ell_{k+3}\text{)}$$

This proves our theorem. ●

The next theorem for Lucas numbers is analogous to Theorem 11.2.

THEOREM 11.8 $\displaystyle\sum_{m=1}^{n} \ell_{2m-1} = \ell_{2m} - 2$

Proof   (by mathematical induction) Let $n = 1$; then $\displaystyle\sum_{m=1}^{1} \ell_{2m-1} = \ell_{2(1)} - 2 = 3 - 2 = 1$.

We assume the proposition is true for $n = k$; that is:

$$\sum_{m=1}^{k} \ell_{2m-1} = \ell_{2k} - 2$$

We must now show that the proposition is true for $n = k + 1$; that is:

$$\sum_{m=1}^{k+1} \ell_{2m-1} = \ell_{2(k+1)} - 2 = \ell_{2k+2} - 2$$

From our assumption, we have:

$$\sum_{m=1}^{k} \ell_{2m-1} + \ell_{2(k+1)-1} = \ell_{2k} - 2 + \ell_{2(k+1)-1}$$

$$\sum_{m=1}^{k+1} \ell_{2m-1} = \ell_{2k} + \ell_{2k+1} - 2 = \ell_{2k+2} - 2$$

This proves our theorem. ●

Because the sum of the first n Lucas numbers has been established (Theorem 11.7) and the sum of the first n odd-numbered terms of the Lucas sequence has been established (Theorem 11.8), it is relatively simple to find the sum of the first n *even-numbered* terms of the Lucas sequence. We use the same method that we used for Fibonacci numbers.

$$\sum_{m=1}^{n} \ell_{2m} = \sum_{m=1}^{2n} \ell_m - \sum_{m=1}^{n} \ell_{2m-1}$$
$$= \ell_{2n+2} - 3 - (\ell_{2n} - 2)$$
$$= \ell_{2n+2} - \ell_{2n} - 3 + 2$$
$$= \ell_{2n+1} - 1$$

This gives us our next theorem, whose proof by mathematical induction is left as an exercise.

THEOREM 11.9 $\displaystyle\sum_{m=1}^{n} \ell_{2m} = \ell_{2n+1} - 1$

The sum of the squares of terms of the Lucas sequence also presents an interesting pattern. Careful inspection of the table of Lucas numbers in Figure 11-11 shows that each entry in the column $\sum \ell_m^2$ is two less than the product of the entry in the same row and the entry in the succeeding row of the column ℓ_m. This implies the following theorem, the proof of which is left as an exercise.

THEOREM 11.10 $\displaystyle\sum_{m=1}^{n} \ell_m^2 = \ell_n \cdot \ell_{n+1} - 2$

To this point, Lucas numbers have appeared as merely a sequence analogous to the Fibonacci sequence (which it is); however, the truly remarkable feature of the Lucas sequence is its *interrelationship* with the Fibonacci sequence. Consider the tables in Figures 11-9 and 11-11 together. For example:

$$f_4 \cdot \ell_4 = 3 \cdot 7 = 21 = f_8$$
$$f_5 \cdot \ell_5 = 5 \cdot 11 = 55 = f_{10}$$
$$f_6 \cdot \ell_6 = 8 \cdot 18 = 144 = f_{12}$$

We might conclude that:

$$f_n \cdot \ell_n = f_{2n} \quad \text{for } n \geq 1 \tag{a}$$

Other interesting relationships between Fibonacci and Lucas numbers evolve:

$$\ell_n = f_{n-1} + f_{n+1} \quad \text{for } n \geq 1 \tag{b}$$
$$5f_n = \ell_{n-1} + \ell_{n+1} \quad \text{for } n \geq 1 \tag{c}$$
$$\ell_n = f_{n+2} - f_{n-2} \tag{d}$$
$$5f_n = \ell_{n+2} - \ell_{n-2} \tag{e}$$

These relations can be proved more easily later. However, if we accept (b) as true, then we can easily prove (c) and (d).

Proof of (c) Because $\ell_n = f_{n-1} + f_{n+1}$, we have:

$$\ell_{n-1} = f_{n-2} + f_n \quad \text{and} \quad \ell_{n+1} = f_n + f_{n+2}$$

Therefore:

$$\begin{aligned}
\ell_{n-1} + \ell_{n+1} &= f_{n-2} + 2f_n + f_{n+2} \\
&= f_n - f_{n-1} + 2f_n + f_n + f_{n+1} \\
&= 4f_n + f_n \\
&= 5f_n \; \bullet
\end{aligned}$$

Proof of (d) $f_{n+2} = f_n + f_{n+1} \quad \text{and} \quad f_{n-2} = f_n - f_{n-1}$

Therefore:

$$f_{n+2} - f_{n-2} = f_n + f_{n+1} - (f_n - f_{n-1}) = f_{n+1} + f_{n-1}$$

Because $\ell_n = f_{n+1} + f_{n-1}$, we get:

$$\ell_n = f_{n+2} - f_{n-2} \; \bullet$$

The reader should attempt other such proofs.

FIBONACCI NUMBERS AND LUCAS NUMBERS IN GEOMETRY

The basic connection of the Fibonacci and Lucas sequences to geometry is through the golden ratio. Consider the ratios of consecutive Fibonacci and Lucas numbers. The table of fractions (Figure 11-12) seems to be approaching ϕ, the golden ratio. Let us investigate this notion. As a review of the golden ratio, consider $\overline{APB}$, with point P partitioning $\overline{AB}$ such that $\dfrac{AB}{AP} = \dfrac{AP}{PB}$ (Figure 11-13).

Let $x = \dfrac{AB}{AP}$. Therefore:

$$x = \frac{AB}{AP} = \frac{AP + PB}{AP} = 1 + \frac{PB}{AP} = 1 + \frac{AP}{AB} = 1 + \frac{1}{x}$$

$\dfrac{f_{n+1}}{f_n}$	$\dfrac{\ell_{n+1}}{\ell_n}$
$\dfrac{1}{1} = 1.000000000$	$\dfrac{3}{1} = 3.000000000$
$\dfrac{2}{1} = 2.000000000$	$\dfrac{4}{3} = 1.333333333$
$\dfrac{3}{2} = 1.500000000$	$\dfrac{7}{4} = 1.750000000$
$\dfrac{5}{3} = 1.666666667$	$\dfrac{11}{7} = 1.571428571$
$\dfrac{8}{5} = 1.600000000$	$\dfrac{18}{11} = 1.636363636$
$\dfrac{13}{8} = 1.625000000$	$\dfrac{29}{18} = 1.611111111$
$\dfrac{21}{13} = 1.615384615$	$\dfrac{47}{29} = 1.620689655$
$\dfrac{34}{21} = 1.619047619$	$\dfrac{76}{47} = 1.617021277$
$\dfrac{55}{34} = 1.617647059$	$\dfrac{123}{76} = 1.618421053$
$\dfrac{89}{55} = 1.618181818$	$\dfrac{199}{123} = 1.617886179$
$\dfrac{144}{89} = 1.617977528$	$\dfrac{322}{199} = 1.618090452$
$\dfrac{233}{144} = 1.618055556$	$\dfrac{521}{322} = 1.618012422$
$\dfrac{377}{233} = 1.618025751$	$\dfrac{843}{521} = 1.618042226$
$\dfrac{610}{377} = 1.618037135$	$\dfrac{1,364}{843} = 1.618030842$
$\dfrac{987}{610} = 1.618032787$	$\dfrac{2,207}{1,364} = 1.618035191$

FIGURE 11-12

FIGURE 11-13

Thus:

$$x = 1 + \frac{1}{x} \quad \text{or} \quad x^2 - x - 1 = 0$$

The roots of this equation are:

$$x_1 = \frac{1 + \sqrt{5}}{2} \approx 1.6180339887 \quad \text{and} \quad x_2 = \frac{1 - \sqrt{5}}{2} \approx -0.6180339887$$

Because we are concerned with lengths of line segments, we will use only the positive root, a. Let a and b represent the roots of the equation $x^2 - x - 1 = 0$:

$$a^2 = a + 1 \tag{I}$$
$$b^2 = b + 1 \tag{II}$$

Multiplying (I) by a^n (where n is an integer), we get:

$$a^{n+2} = a^{n+1} + a^n \tag{III}$$

Multiplying (II) by b^n (where n is an integer), we get:

$$b^{n+2} = b^{n+1} + b^n \tag{IV}$$

Subtracting (IV) from (III) gives us:

$$a^{n+2} - b^{n+2} = (a^{n+1} - b^{n+1}) + (a^n - b^n)$$

Now dividing by $a - b = \sqrt{5}$ (thus nonzero) gives us:

$$\frac{a^{n+2} - b^{n+2}}{a - b} = \frac{a^{n+1} - b^{n+1}}{a - b} + \frac{a^n - b^n}{a - b}$$

Let $t_n = \dfrac{a^n - b^n}{a - b}$. Then $t_{n+2} = t_{n+1} + t_n$, the same as the Fibonacci sequence definition.

All that remains to be shown in order to be able to establish t_n as the nth Fibonacci number, f_n, is that $t_1 = 1$ and $t_2 = 1$. We have:

$$t_1 = \frac{a^1 - b^1}{a - b} = 1$$
$$t_2 = \frac{a^2 - b^2}{a - b} = \frac{(a - b)(a + b)}{a - b} = \frac{(\sqrt{5})(1)}{\sqrt{5}} = 1$$

Therefore:

$$f_n = \frac{a^n - b^n}{a - b} \quad \text{where } a = \frac{1 + \sqrt{5}}{2}, b = \frac{1 - \sqrt{5}}{2}, \text{ and } n = 1, 2, 3, \ldots$$

This is how a Fibonacci number is expressed in *Binet form*, that is, in terms of non-Fibonacci numbers.

Example   Find f_6.

Solution
$$f_6 = \frac{a^6 - b^6}{a - b} = \frac{(a^3 - b^3)(a^3 + b^3)}{a - b} = \frac{(a - b)(a^2 + ab + b^2)(a^3 + b^3)}{a - b}$$
$$= (a^2 + ab + b^2)(a^2 - ab + b^2)(a + b)$$
$$= (2)(4)(1) = 8 \; \bullet$$

Now add (III) and (IV) from page 232:

$$a^{n+2} + b^{n+2} = (a^{n+1} + b^{n+1}) + (a^n + b^n)$$

Let $w_n = a^n + b^n$. Therefore:

$$w_{n+2} = w_{n+1} + w_n$$

To further inspect the sequence w_n, consider w_1 and w_2:

$$w_1 = a^1 + b^1 = \frac{1 + \sqrt{5}}{2} + \frac{1 - \sqrt{5}}{2} = 1$$
$$w_2 = a^2 + b^2 = \left(\frac{1 + \sqrt{5}}{2}\right)^2 + \left(\frac{1 - \sqrt{5}}{2}\right)^2$$
$$= \frac{6 + 2\sqrt{5}}{4} + \frac{6 - 2\sqrt{5}}{4} = 3$$

Because $w_1 = 1$, $w_2 = 3$, and $w_{n+2} = w_{n+1} + w_n$, w_n is the nth Lucas number:

$$\ell_n = a^n + b^n \quad \text{where } a = \frac{1 + \sqrt{5}}{2}, b = \frac{1 - \sqrt{5}}{2}, \text{ and } n = 1, 2, 3, \ldots$$

This is how a Lucas number is expressed in Binet form.

Example   Prove that $f_{2n} = f_n \cdot \ell_n$ (equation (a) on page 229).

Solution   Because $f_n = \dfrac{a^n - b^n}{a - b}$ and $\ell_n = a^n + b^n$, we have:

$$f_n \cdot \ell_n = \left(\frac{a^n - b^n}{a - b}\right) \cdot (a^n + b^n)$$
$$= \frac{a^{2n} - b^{2n}}{a - b} = f_{2n} \; \bullet$$

Returning to the table of fractions in Figure 11-12, we now consider $\dfrac{f_{n+1}}{f_n}$ and $\dfrac{\ell_{n+1}}{\ell_n}$.

$$\lim_{n \to \infty} \frac{f_{n+1}}{f_n} = \frac{\dfrac{a^{n+1} - b^{n+1}}{a - b}}{\dfrac{a^n - b^n}{a - b}} = \frac{a^{n+1} - b^{n+1}}{a^n - b^n}$$

Now we divide all terms by a^n:

$$\lim_{n\to\infty}\frac{f_{n+1}}{f_n} = \frac{a - \dfrac{b^{n+1}}{a^n}}{1 - \dfrac{b^n}{a^n}}$$

Because $\lim_{n\to\infty} b^n = 0$, we have:

$$\lim_{n\to\infty}\frac{f_{n+1}}{f_n} = a = \frac{1+\sqrt{5}}{2} = \phi \quad \text{(the golden ratio)}$$

This result justifies the conjecture we made at the start of this section.

The proof that $\lim_{n\to\infty}\dfrac{\ell_{n+1}}{\ell_n} = \phi$ may be carried out in a similar way. We now have the following two theorems.

THEOREM 11.11 $\quad \lim_{n\to\infty}\dfrac{f_{n+1}}{f_n} = \phi$

THEOREM 11.12 $\quad \lim_{n\to\infty}\dfrac{\ell_{n+1}}{f_n} = \phi$

It is interesting to consider an alternate proof of Theorem 11.11.

P roof   Let $x = \lim_{n\to\infty}\dfrac{f_{n+1}}{f_n}$. The definition of f_n enables us to get:

$$f_{n+1} = f_n + f_{n-1}$$

Therefore:

$$x = \lim_{n\to\infty}\frac{f_n + f_{n+1}}{f_n}$$
$$= \lim_{n\to\infty}\left(1 + \frac{f_{n-1}}{f_n}\right)$$
$$= \lim_{n\to\infty}1 + \lim_{n\to\infty}\frac{f_{n-1}}{f_n}$$

We have:

$$\lim_{n\to\infty}\frac{f_n}{f_{n-1}} = \lim_{n\to\infty}\frac{f_{n+1}}{f_n}$$

Therefore:

$$x = 1 + \frac{1}{x} \Rightarrow x^2 - x - 1 = 0 \quad \text{and} \quad x = \phi \; \bullet$$

The Binet form of f_n provides us with a simple way to establish equation (b) on page 229.

$$\begin{aligned}
\ell_n &= f_{n-1} + f_{n+1} \\
&= \frac{a^{n-1} - b^{n-1}}{a - b} + \frac{a^{n+1} - b^{n+1}}{a - b} \\
&= \frac{\dfrac{a^n}{a} - \dfrac{b^n}{b} + a \cdot a^n - b \cdot b^n}{a - b} \\
&= \frac{a^n\left(a + \dfrac{1}{a}\right) - b^n\left(b + \dfrac{1}{b}\right)}{a - b}
\end{aligned}$$

Because $a = \dfrac{1}{b}$ and $b = \dfrac{1}{a}$, we have:

$$\begin{aligned}
\ell_n &= a^n(a + b) - b^n(a + b) \\
&= a^n(1) - b^n(1) \\
&= a^n - b^n
\end{aligned}$$

We now return to the powers of ϕ, which originally spurred our discussion of the Fibonacci numbers (see Figure 11-7). With our knowledge of these numbers, we can state a general term for powers of ϕ:

$$\phi^n = f_n\phi + f_{n-1}$$

We now have two clear connections between the golden ratio, ϕ, and Fibonacci numbers.

THE GOLDEN RECTANGLE REVISITED

We continue with our discussion of golden rectangle *ABCD*. We established in Figure 11–6 that when a square is constructed internally (as in Figure 11-14), if $AF = 1$, then $FD = \dfrac{1}{\phi}$ and $AD = 1 + \dfrac{1}{\phi} = \phi$. Thus rectangle *CDFE* in Figure 11-14 has dimensions $FD = \dfrac{1}{\phi}$ and $CD = 1$ and is also a golden rectangle.

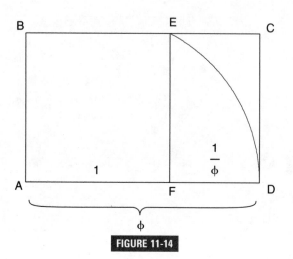

FIGURE 11-14

We continue the process of constructing an internal square. In golden rectangle *CDFE*, we construct square *DFGH* (Figure 11-15). We find that $CH = 1 - \frac{1}{\phi} = \frac{1}{\phi^2}$, thereby establishing rectangle *CHGE* as a golden rectangle. Continuing this scheme, we construct square *CHKJ* in golden rectangle *CHGE* and find that $EJ = \frac{1}{\phi} - \frac{1}{\phi^2} = \frac{\phi - 1}{\phi^2} = \frac{\frac{1}{\phi}}{\phi^2} = \frac{1}{\phi^3}$. (Note: We showed earlier that $\phi - \frac{1}{\phi} = 1$). Once again, we have a new golden rectangle, this time rectangle *EJKG*. By continuing this process, we get golden rectangle *GKML*, golden rectangle *NMKR*, golden rectangle *MNST*, and so on.

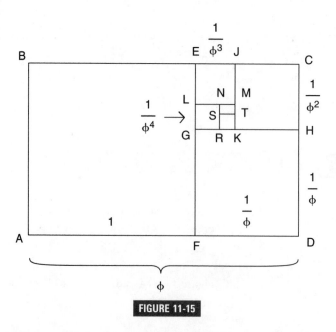

FIGURE 11-15

Suppose we now draw the following quadrants (quarter circles) (see Figure 11-16):

center *E*, radius *EB*

center *G*, radius *GF*

center *K*, radius *KH*

center *M*, radius *MJ*

center *N*, radius *NL*

center *S*, radius *SR*

.
.
.

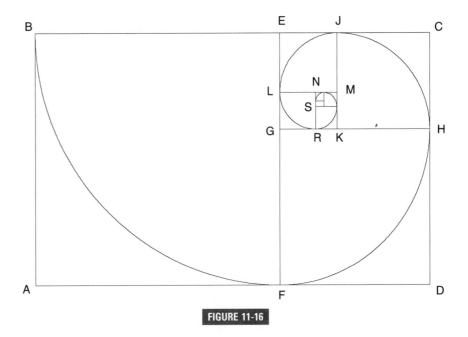

FIGURE 11-16

The result is a logarithmic spiral. We also notice that the centers of these squares lie in another logarithmic spiral (Figure 11-17).

The spiral in Figure 11-16 seems to converge at a point in rectangle *ABCD*. This point is at the intersection, point *P*, of $\overline{AC}$ and $\overline{ED}$ (Figure 11-18). Consider once again golden rectangle *ABCD* (Figure 11-18). Earlier we established that square *ABEF* determines another golden rectangle, *CEFD*. Because all golden rectangles have the same shape, rectangle *ABCD* is similar to rectangle *CEFD*. This implies that $\triangle ECD \sim \triangle CDA$. Therefore $\angle CED \cong \angle DCA$ and $\angle DCA$ is complementary to $\angle ECA$. Therefore $\angle CED$ is complementary to $\angle ECA$. Thus $\angle EPC$ must be a right angle, which tells us that $\overline{AC} \perp \overline{ED}$.

If the width of one rectangle is the length of the other and the rectangles are similar, then the rectangles are said to be *reciprocal rectangles*. In Figure 11-18, we see that rectangle *ABCD* and rectangle *CEFD* are reciprocal rectangles.

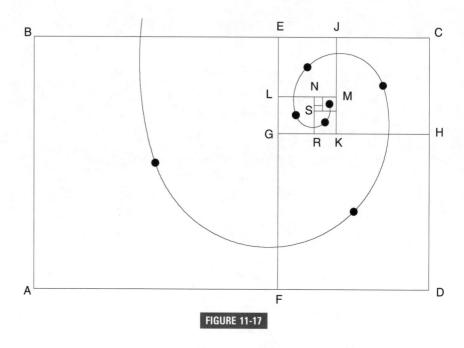

FIGURE 11-17

Furthermore, we see that reciprocal rectangles have corresponding diagonals perpendicular. In the same way as before, we can prove that rectangles *CEFD* and *CEGH* are reciprocal rectangles, with diagonals $\overline{ED}$ and $\overline{CG}$ perpendicular at point *P*. We may extend this proof to each pair of consecutive golden rectangles in Figure 11-16. Clearly point *P* ought to be the limit of the spiral in Figure 11-16.

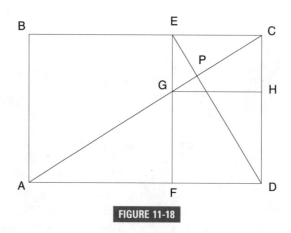

FIGURE 11-18

THEOREM 11.13 Reciprocal rectangles have perpendicular diagonals.

Proof The case in which one rectangle is in the interior of the other rectangle was proved earlier. Therefore we will consider only the case in which the two reciprocal rectangles share no common interior region (Figure 11-19).

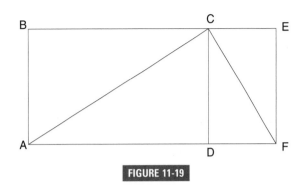

FIGURE 11-19

Rectangles *ABCD* and *CEFD* in Figure 11-19 are similar. Therefore △*CDA* ~ △*FDC* and ∠*FCD* ≅ ∠*CAD*. However, ∠*CAD* is complementary to ∠*DCA*. Therefore ∠*FCD* is complementary to ∠*DCA*. Thus $\overline{AC} \perp \overline{CF}$. ●

THEOREM 11.14   If two rectangles have one pair of corresponding diagonals perpendicular and a width of one is the length of the other, then the rectangles are reciprocal.

℗roof   Rectangles *ABCD* and *CEFD* have diagonals $\overline{AC} \perp \overline{ED}$ at point *P* (Figure 11-20). We must show that the two rectangles are similar.

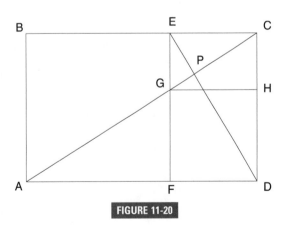

FIGURE 11-20

Both ∠*CED* and ∠*DCA* are complementary to ∠*ECA*. Therefore ∠*CED* ≅ ∠*DCA*. This enables us to establish that △*CED* ~ △*DCA*, whereupon it follows that the rectangles are similar (because the corresponding sides are proportional). ●

Theorem 11.14 has interesting applications. For one thing, it provides us with another way to construct consecutive golden rectangles. We could simply begin with golden rectangle *ABCD*, construct a perpendicular from point *D* to $\overline{AC}$, and from its intersection *E* with $\overline{BC}$ construct a perpendicular to $\overline{AD}$ to complete the second golden rectangle. This process can be repeated indefinitely.

THE GOLDEN TRIANGLE

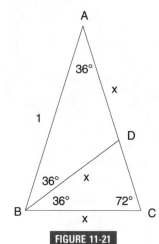

FIGURE 11-21

Just as we "admired" the golden rectangle, so too have we a triangle worthy of admiration. It is built on the golden ratio and has many wonderful properties, somewhat analogous to those of the golden rectangle. We begin with an isosceles triangle whose vertex angle has measure 36° and construct the bisector $\overline{BD}$ of $\angle ABC$ (Figure 11-21). We find that $m\angle DBC = 36°$. Therefore $\triangle ABC \sim \triangle BCD$. Let $AD = x$ and $AB = 1$. Because $\triangle ADB$ and $\triangle DBC$ are isosceles, $BC = BD = AD = x$. From the similarity of triangles ABC and BCD:

$$\frac{1}{x} = \frac{x}{1-x}$$

This gives us:

$$x^2 + x - 1 = 0 \quad \text{and} \quad x = \frac{\sqrt{5}-1}{2}$$

(We disregard the negative root because we are dealing with length.)

Recall that $\dfrac{\sqrt{5}-1}{2} = \dfrac{1}{\phi}$. The ratio $\dfrac{\text{side}}{\text{base}}$ for $\triangle ABC$ is $\dfrac{1}{x} = \phi$. We therefore call $\triangle ABC$ a *golden triangle*.

By taking consecutive angle bisectors $\overline{BD}, \overline{CE}, \overline{DF}, \overline{EG},$ and $\overline{FH}$ of a base angle of each newly formed 36-72-72 triangle, we get a series of golden triangles (see Figure 11-22). These golden triangles (with angles of measure 36°, 72°, 72°) are $\triangle ABC, \triangle BCD, \triangle CDE, \triangle DEF, \triangle EFG,$ and $\triangle FGH$. Had space permitted we could have continued to draw angle bisectors and thereby generate more golden triangles.

Our study of the golden triangle parallels that of the golden rectangle. We begin by letting $HG = 1$ (Figure 11-22). Because the ratio $\dfrac{\text{side}}{\text{base}}$ of a golden triangle is ϕ, we find that for golden triangle FGH:

$$\frac{GF}{HG} = \frac{\phi}{1} \quad \text{or} \quad \frac{GF}{1} = \frac{\phi}{1}$$

Therefore $GF = \phi$.

Similarly, for golden $\triangle EFG$, $\dfrac{FE}{GF} = \dfrac{\phi}{1}$. But $GF = \phi$, so:

$$FE = \phi^2$$

In golden $\triangle DEF$, $\dfrac{ED}{FE} = \dfrac{\phi}{1}$. But $FE = \phi^2$, so:

$$ED = \phi^3$$

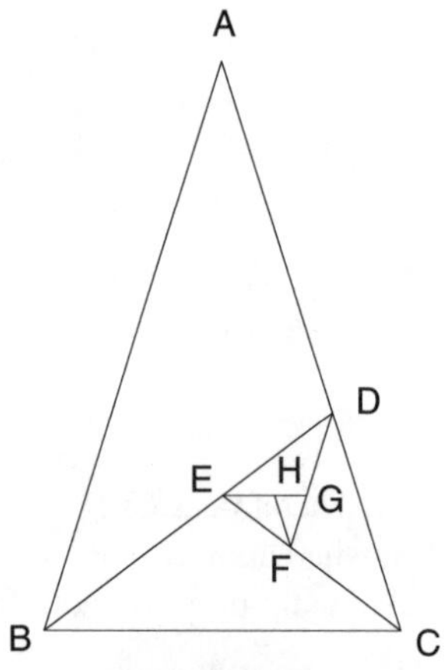

FIGURE 11-22

Again, for $\triangle CDE$, $\dfrac{DC}{ED} = \dfrac{\phi}{1}$. But $ED = \phi^3$, so:

$$DC = \phi^4$$

For $\triangle BCD$, $\dfrac{CB}{DC} = \dfrac{\phi}{1}$. But $DC = \phi^4$, so:

$$CB = \phi^5$$

Finally, for $\triangle ABC$, $\dfrac{BA}{CB} = \dfrac{\phi}{1}$. But $CB = \phi^5$, so:

$$BA = \phi^6$$

Using our knowledge of powers of ϕ (developed earlier), we can summarize these results as follows:

$$HG = \phi^0 = 0\phi + 1 = f_0\phi + f_{-1}$$
$$GF = \phi^1 = 1\phi + 0 = f_1\phi + f_0$$
$$FE = \phi^2 = 1\phi + 1 = f_2\phi + f_1$$
$$ED = \phi^3 = 2\phi + 1 = f_3\phi + f_2$$
$$DC = \phi^4 = 3\phi + 2 = f_4\phi + f_3$$
$$CB = \phi^5 = 5\phi + 3 = f_5\phi + f_4$$
$$BA = \phi^6 = 8\phi + 5 = f_6\phi + f_5$$

As we did with the golden rectangle, we can generate a logarithmic spiral by drawing arcs to join the vertex angle vertices of consecutive golden triangles (see Figure 11-23).

That is, we draw circular arcs as follows:

$\overgroup{AB}$ (circle center at point D)

$\overgroup{BC}$ (circle center at point E)

$\overgroup{CD}$ (circle center at point F)

$\overgroup{DE}$ (circle center at point G)

$\overgroup{EF}$ (circle center at point H)

$\overgroup{FG}$ (circle center at point J)

$\vdots$

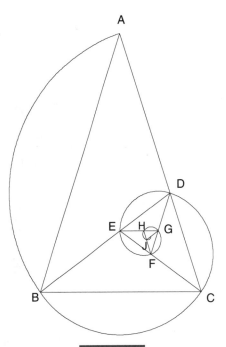

FIGURE 11-23

Many other truly fascinating relationships emanate from the golden ratio. After you have been exposed to the golden triangle, the next logical place to turn for more applications is the regular pentagon and regular pentagram (the five-pointed star), which are essentially composed of many golden triangles. We offer a number of these and other relationships as exercises for you to work on with the hope that they will entice you to further study the golden ratio and Fibonacci numbers. Along the way you will certainly come to appreciate the close connection between geometry and other branches of mathematics.

EXERCISE**S**

Verify each of the relationships in exercises 1–4.

1. a. $\dfrac{f_{12}}{f_6} = f_5 + f_7$ b. $\dfrac{\ell_{12}}{\ell_4} = \ell_8 - 1$

2. a. $\dfrac{f_{10}}{f_5} = f_4 + f_6$ b. $\dfrac{\ell_{15}}{\ell_5} = \ell_{10} + 1$

3. a. $f_{11} = f_5^2 + f_6^2$ b. $\dfrac{\ell_9}{\ell_3} = \ell_6 + 1$

4. a. $f_9^2 - 4 \cdot f_8 \cdot f_7 = f_6^2$ b. $\dfrac{\ell_6}{\ell_2} = \ell_4 - 1$

Using the table in Figure 11-9, fill in each missing term in exercises 5–10.

5. $f_n^2 + f_{n+1}^2 = $ _____

6. $f_{2n} = f_{n+1}^2 - $ _____

7. $f_{-n} = $ _____ $\cdot f_n$ (You might want to extend the table of Figure 11-9.)

8. $f_{2n} \div $ _____ $= f_{n-1} + f_{n+1}$

9. $f_{n+3}^2 - f_{n-3}^2 = f_6 \cdot $ _____

10. $\displaystyle\sum_{m=1}^{6} f_m = $ _____ $- 1$

11. Represent $\displaystyle\sum_{m=1}^{5} f_m^2$ geometrically.

12. Represent $\displaystyle\sum_{m=1}^{8} f_m^2$ geometrically.

Express each sum in exercises 13–16 in terms of other Fibonacci numbers.

13. $\displaystyle\sum_{m=1}^{n} f_{4m}$
14. $\displaystyle\sum_{m=1}^{n} f_{4m-1}$

15. $\displaystyle\sum_{m=1}^{n} f_{4m-2}$
16. $\displaystyle\sum_{m=1}^{n} f_{4m-3}$

Prove each of exercises 17–21 by mathematical induction.

17. $\displaystyle\sum_{m=1}^{n} f_{2m} = f_{2n+1} - 1$

18. $\displaystyle\sum_{m=1}^{n} f_m^2 = f_n \cdot f_{n+1}$

19. $f_n^2 - f_{n+2} \cdot f_{n-2} = (-1)^n$

20. $f_{2n} + f_{n-1}^2 = f_{n+1}^2$

21. $f_{2n} = f_n \cdot f_{n-1} + f_n \cdot f_{n+1}$

22–25. Prove by mathematical induction each of the results obtained in exercises 13–16.

26. Show that the sum of the sums of exercises 13–16 equals $\displaystyle\sum_{m=1}^{4n} f_m$.

Using the tables in Figures 11-9 and 11-11, fill in the missing term in each of exercises 27–30.

27. $5f_{2n+1} = \ell_{n+1}^2 +$ _____

28. $\ell_{4n} + 2 =$ _____

29. $5f_{2n}^2 =$ _____ $- 2$

30. $5f_{2n}^2 =$ _____ $- 4$

31. Represent $\displaystyle\sum_{m=1}^{5} \ell_m^2$ geometrically.

32. Represent $\displaystyle\sum_{m=1}^{8} \ell_m^2$ geometrically.

Express each of the sums in exercises 33–36 in terms of Fibonacci numbers.

33. $\displaystyle\sum_{m=1}^{n} \ell_{4m}$
34. $\displaystyle\sum_{m=1}^{n} \ell_{4m-1}$

35. $\displaystyle\sum_{m=1}^{n} \ell_{4m-2}$
36. $\displaystyle\sum_{m=1}^{n} \ell_{4m-3}$

37. Prove Theorem 11.9 by mathematical induction.

38. Prove Theorem 11.10 by mathematical induction.

39. Show by example how the first 25 natural numbers may be expressed as the sum of distinct Lucas numbers.

40. Show how any natural number may be expressed as the sum of distinct Fibonacci numbers (e.g., $15 = f_3 + f_7 = f_3 + f_5 + f_6$, etc., or $45 = f_4 + f_6 + f_9$). First represent the numbers from 1 to 100 in terms of Fibonacci numbers; then try to prove that any natural number can be expressed in terms of Fibonacci numbers. Can all natural numbers be represented in terms of Fibonacci numbers if any *one* Fibonacci number is missing? If any *two* Fibonacci numbers are missing? Show by examples that if f_1 is missing, then any natural number can be expressed in terms of Fibonacci numbers (excluding f_1) in exactly one way if no two consecutive terms (f_k and f_{k+1}) are chosen (e.g., $17 = f_2 + f_4 + f_7$).

41. Set up a table of differences of the terms of the Fibonacci sequence. You might want to start from the table in Figure 11-24 and then expand it. Notice the patterns!

5	−3	2	−1	1	0	1	1	2	3	5	8
	−8	5	−3	2	−1	1	0	1	1	2	3
		13	−8	5	−3	2	−1	1	0	1	1
			−21	13	−8	5	−3	2	−1	1	0
				34	−21	13	−8	5	−3	2	−1
					−55	34	−21	13	−8	5	−3
						89	−55	34	−21	13	−8
							−144	89	−55	34	−21

FIGURE 11-24

42. Set up a table of differences for the Lucas numbers as you did for Fibonacci numbers in exercise 41.

43. Consider the divisibility of Fibonacci numbers. Note that $f_5 \mid f_{10}$ (which reads "f_5 divides f_{10}") and $f_5 \mid f_{15}$; also $f_4 \mid f_8, f_4 \mid f_{12}, f_4 \mid f_{16}$, and so on. Find a general pattern for divisibility of Fibonacci numbers and justify your conclusion.

44. Divide each of $f_1, f_2, f_3, \ldots, f_{31}$ by 7 and inspect the remainders. Now divide each of $f_1, f_2, f_3, \ldots, f_{31}$ by 5 and inspect the remainders. What conclusions can you draw regarding the remainders? Which natural numbers divide Fibonacci numbers? Try divisions by other natural numbers and establish various patterns.

45. The Pascal triangle is shown in Figure 11-25. See if you can find sums of numbers that generate the Fibonacci sequence. Can you justify this?

Use the Binet form of f_n and ℓ_n to find the value of each expression in exercises 46–49.

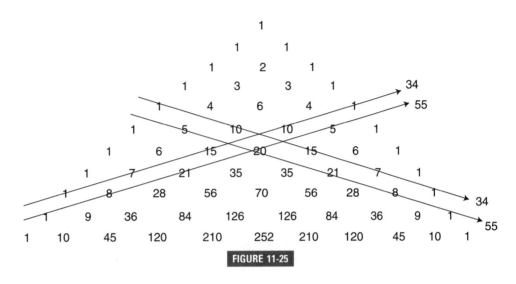

FIGURE 11-25

46. f_7

47. ℓ_7

48. ℓ_9

49. f_9

Verify each of the equalities in exercises 50–55 with the Binet form of f_n and ℓ_n.

50. $\ell_n = f_{n-1} + f_{n+1}$

51. $5f_n = \ell_{n-1} + \ell_{n+1}$

52. $\ell_n = f_{n+2} - f_{n-2}$

53. $5f_n = \ell_{n+2} + \ell_{n-2}$

54. $5f_n^2 = \ell_n^2 - 4(-1)^n$

55. $5f_{2n+1} = \ell_{n+1}^2 + \ell_n^2$

56. Prove by mathematical induction that $f_n = \dfrac{a^n - b^n}{a - b}$, where $a = \dfrac{1 + \sqrt{5}}{2}$, $b = \dfrac{1 - \sqrt{5}}{2}$, and $n = 1, 2, 3, \ldots$.

57. Prove by mathematical induction that $\ell_n = a^n + b^n$, where a, b, and n are as defined in exercise 56.

58. Consider the quotient $\dfrac{1}{1 - x - x^2} = f_1 + f_2 x + f_3 x^2 + \cdots + f_n x^{n-1} + \cdots,$ where f_n is the nth Fibonacci number. Justify why this relationship occurs.

59. Consider the powers of $a = \phi$ (the golden ratio), that is:

$$a = \frac{1 + \sqrt{5}}{2}$$

$$a^2 = \frac{6 + 2\sqrt{5}}{4} = \frac{3 + \sqrt{5}}{2}$$

$$a^3 = \frac{8 + 4\sqrt{5}}{4} = \frac{4 + 2\sqrt{5}}{2}$$

$$a^4 = \frac{14 + 6\sqrt{5}}{4} = \frac{7 + 3\sqrt{5}}{2}$$

$$a^5 = \frac{22 + 10\sqrt{5}}{4} = \frac{11 + 5\sqrt{5}}{2}$$

$$a^6 = \frac{36 + 16\sqrt{5}}{4} = \frac{18 + 8\sqrt{5}}{2}$$

Discuss this result as well as other representations of powers of $a = \phi$, such as:

$$a^2 = \frac{3 + \sqrt{5}}{2} = \frac{1 + \sqrt{5}}{2} + 1 = a + 1$$

$$a^3 = 2a + 1$$

$$a^4 = 3a + 2$$

$$\vdots$$

60. Establish some additional connections between the golden ratio and Fibonacci and Lucas numbers.

61. Prove that the lengths of the sides of a triangle can never be three consecutive Fibonacci numbers or three consecutive Lucas numbers.

62. Prove that if points P, Q, R, and S partition each of the sides of square $ABCD$ into the golden ratio, as shown in Figure 11-26, then quadrilateral $PQRS$ is a golden rectangle.

63. Identify the golden section on the regular pentagram in as many places as you can.

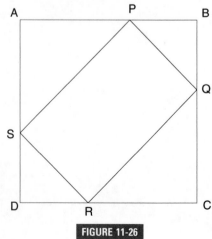

FIGURE 11-26

64. Identify the golden section on the regular pentagon in as many places as you can.

65. Prove that the following construction partitions $\overline{AB}$ into the golden ratio: Construct a perpendicular $\overline{CB}$ to $\overline{AB}$ at point B so that $CB = \frac{1}{2}(AB)$. Draw $\overline{AC}$ (see Figure 11-27). Then draw a circle with center C and radius $\overline{CB}$ to intersect $\overline{AC}$ at point F. The point, P, at which the circle with center A and radius $\overline{AF}$ intersects $\overline{AB}$ is the point at which $\overline{AB}$ is partitioned into the golden section; that is, $\dfrac{AB}{AP} = \dfrac{AP}{PB}$.

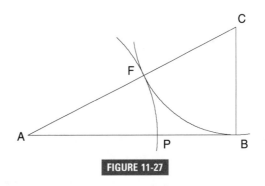

FIGURE 11-27

66. Prove that in Figure 11-22 area $\triangle BCD$: area $\triangle ABD$: area $\triangle ABC =$ $1 : \phi : \phi^2$.

67. Show that $\sin 18° = \dfrac{1}{2\phi}$.

68. Find $\cos 27°$ in terms of ϕ.

69. Prove that the bisector of the vertex angle of a golden triangle partitions each of the other angle bisectors into the golden ratio.

70. Prove that the following construction partitions $\overline{QB}$ into the golden ratio: Begin with square $ABCD$ (Figure 11-28) and construct a semicircle internally on $\overline{AB}$. With M the midpoint of $\overline{AB}$, draw $\overline{CM}$ and $\overline{DM}$ to intersect the semicircle at points E and F, respectively. From point E, construct a perpendicular to $\overline{AB}$ at point P. From point F, construct a perpendicular to $\overline{AB}$ at point Q. $\dfrac{PQ}{BP} = \phi$. (The dashed lines in Figure 11-28 serve as a hint.)

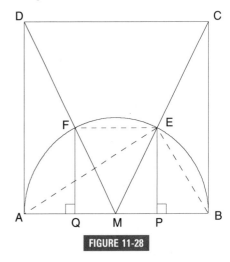

FIGURE 11-28

71. Prove that the following construction partitions $\overline{MN}$ into the golden ratio: Begin with square $ABCD$, with M the midpoint of $\overline{CD}$. Circle I is inscribed in $\triangle AMB$ and is tangent to $\overline{AM}$, $\overline{BM}$, and $\overline{AB}$ at points E, F, and N, respectively (Figure 11-29). Also, $\overline{MN}$ intersects circle I at point P. We then have point P partitioning $\overline{MN}$ into the golden ratio; that is, $\dfrac{MP}{PN} = \dfrac{PN}{MN}$.

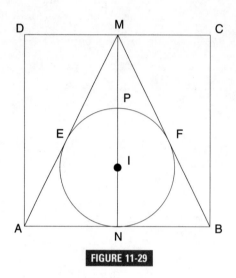

FIGURE 11-29

72. Show that the lengths of the medians of the base angles of consecutive golden triangles form a Fibonacci sequence.

73. Prove that the area of a regular pentagon with side length s equals $\dfrac{5s^2\phi}{4\sqrt{3-\phi}}$.

INDEX